CITIZEN CYBORG

CITIZEN CYBORG

Why Democratic Societies Must Respond to the Redesigned Human of the Future

James Hughes

A Member of the Perseus Books Group

Copyright © 2004 by James Hughes

Published in 2004 in the United States of America by Westview Press.

Find us on the world wide web at www.westviewpress.com

Westview Press books are available at special discounts for bulk purchases in the United States by corporations, institutions, and other organizations. For more information, please contact the Special Markets Department at the Perseus Books Group, 11 Cambridge Center, Cambridge, MA 02142, or call (800) 255-1514 or (617) 252-5298, or e-mail special.markets@perseusbooks.com.

Cataloging-in-Publication data is available from the Library of Congress.
ISBN-10 0-8133-4198-1 (hc.)
ISBN-13 978-0-8133-4198-9 (hc.)

For Althea and Tristan

May your ride on the waves of change be long and full of adventure

Contents

━━━━━━━━━━━━━━ SECTION III ━━━━━━━━━━━━━━

FREEDOM AND EQUALITY AMONG THE CYBORGS

Acknowledgments

First and foremost I would like to thank my excellent agent, Ted Weinstein, who patiently guided me through the process of crafting my decade-old embryonic book into the perfect brain child I always wanted.

My colleagues back at the MacLean Center for Clinical Ethics at the University of Chicago—Mark Siegler, John Lantos, Carol Stocking, Ann Dudley Goldblatt and the rest—provided a very congenial working environment for the six years that I worked and taught with them. This book was first conceived and outlined there and I wrote the book's seminal essay, "Embracing Change with All Four Arms," for a seminar on Health and Humanities that I ran for the Center.

From the MacLean Center the line of causation runs to Calixto Machado, the coordinator of the International Network for the Definition of Death, whose ongoing friendship and collaboration on the work of the Network led to my writing on brain death. My brain death essays led, in turn, to an invitation from Alcor to present my thoughts to them on "The Future of Death" in 2001, which led to a publication in the *Journal of Transhumanism*. The editor of *JOT* (now *JET*), philosopher Mark Walker of the University of Toronto, became a friend and coconspirator and has contributed many acknowledged and unacknowledged ideas to this book. Without the support, friendship and feedback of the founder and chair of the World Transhumanist Association, Nick Bostrom, I would never have thrown myself into transhumanism or found the passion necessary to complete this book.

More generally I have benefited tremendously from the ongoing sharpening and refining of ideas in daily intercourse with the participants of the discussion lists of the World Transhumanist Association. The book has benefited in particular from many fine insights from Dale Carrico and my co-conspirators at the Cyborg Democracy blog. I look forward to having this book ruthlessly dissected on the transhumanist lists and in the blogosphere.

I owe a debt of thanks to those who have encouraged and supported Changesurfer Radio and my Change Surfing column, both of which have forced me to refine my message. Richard Sherman, John Schwenk and the wonderful radical community around WHUS 91.7 FM at the University of Connecticut have supported Changesurfer Radio since 1997. Simon Smith and George Dvorsky of Betterhumans.com have been patient and persistent editors of my Change Surfing columns, often deserving a byline.

Just as the project was conceived at the University of Chicago ethics center, it owes its successful birth to the support and congeniality of Trinity College, where I have worked since 1999. My friend and colleague Kent Smith has been extraordinarily understanding about my distractedness and occasional absences during this last year.

Finally, my wife and children have tolerated many nights and weekends when I have been physically absent, and many more when I have been mentally absent. Thanks for your patience and I'll try to make it up to you in the centuries to come.

Introduction

This book began in the spring of 1986 on a bus rolling through the small, crooked, immaculate, beautiful streets of Kyoto. I had two books in my lap—*Algeny,* a shrill, alarmist attack on gene technologies by Jeremy Rifkin, and *What Sort of People Should There Be?,* a cool, moderate and thorough defense of genetic engineering and virtual reality by Oxford don Jonathan Glover. Reading the two books I realized that here lay the outer boundaries of twenty-first-century biopolitics.

That winter my father, an insurance actuary, had called to see if I could find out how many organ transplants were being done in Japan. He needed to write a policy to cover employees who might need a transplant while living in Japan. While researching the question I became fascinated with the Japanese debate about brain death and abortion. The American right-to-life movement opposed both abortion and the recognition of brain death on the grounds that human life was precious from conception to heart death. Many American secular humanists believed abortion and euthanasia of the brain dead were acceptable because self-awareness gave human life meaning. The Japanese leaned in a different direction altogether. The Japanese had little problem with abortion. In fact, Japanese women often turned to Buddhist monks to perform rituals to clean up the bad karma they might incur from abortion.

But the Japanese had a much harder time swallowing the idea of brain death. Without a brain death law the Japanese couldn't permit hearts, livers or pancreases to be transplanted from brain dead bodies on respirators. Taking a vital organ out of someone in a permanent coma would be murder unless they were already "dead." The would-be Japanese transplantors were lobbying hard. The debate was so interesting I resolved to give up my studies of Buddhist history and politics, and go back to the United States to start graduate work comparing the Japanese and American personhood debates in abortion and brain death.

Sitting on the bus that day, illuminated by Rifkin and Glover, I came to a second realization: The human race's use of genetic engineering to evolve beyond our current limitations would be a central political issue of the next century. Just as in abortion and brain death, the key issue in genetic engineering was whether it is more important that we remain "human" or that we are "persons." Is there anything we must preserve about *Homo sapiens* DNA or "human nature"? The last two decades have added new tools to transcend our limitations, such as nanotechnology, but the basic question remains the same.

In the twenty-first century the convergence of artificial intelligence, nanotechnology and genetic engineering will allow human beings to achieve things previously imagined only in science fiction. Life spans will extend well beyond a century. Our senses and cognition will be enhanced. We will gain control over our emotions and memory. We will merge with machines, and machines will become more like humans. These technologies will allow us to evolve into varieties of "posthumans" and usher us into a "transhuman" era and society.

This book argues that transhuman technologies, technologies that push the boundaries of humanness, can radically improve our quality of life, and that we have a fundamental right to use them to control our bodies and minds. But to ensure these benefits we need to democratically regulate these technologies and make them equally available in free societies. Becoming more than human can improve all our lives, but only new forms of transhuman citizenship and democracy will make us freer, more equal and more united.

A lot of people are understandably frightened by the idea of a society in which unenhanced humans will need to coexist with humans who are smarter, faster and more able, not to mention robots and enhanced animals. How can democracy embrace transhuman diversity when the gap between the rich and poor continues to widen, and there is so much discrimination against women and people of color? Professional fear-mongers spin fanciful scenarios based on classic horror movies—*The Island of Dr. Moreau, Frankenstein, Brave New World, The Boys from Brazil, Gattaca, The Terminator.* We can't genetically engineer tomatoes, the argument goes, because ants that ate genemod tomatoes in the movies went on to eat Cleveland.

The "bioLuddite" opposition to genetic engineering, nanotechnology and artificial intelligence, slowly building and networking since the 1960s, picked up where the anti-industrialization Luddites left off in the nineteenth century. While Luddites believed that defending workers' rights required a ban on the automation of work, the bioLuddites believe genetic engineering and human enhancement technologies cannot be used safely, and must be summarily banned. Decades of patient effort by anti-technology activists, such as Jeremy Rifkin, have created broad left-to-right bioLuddite coalitions of religious conservatives, disability rights supporters, feminists and environmental and anti-corporate activists opposed to genetically engineered food, embryonic stem cell cloning, nanotechnology and now human enhancement. At root the bioLuddites are also rejecting liberal democracy, science and modernity. They have given up on the idea of progress guided by human reason, and, afraid of the radical choices and diversity of a transhuman future, are reasserting mystical theories of natural law and order. Whether secular bioethicists, ecomystic Greens or religious fundamentalists, the bioLuddites insist that there are clear and obvious boundaries to what people should be allowed to do with their own bodies, and that no one should be allowed to become something more than human.

Fear-mongering about human enhancement hit a high-water mark in 2001 when U.S. president George W. Bush appointed the conservative philosopher Leon Kass to head his new President's Council on Bioethics (PCB). Kass's appointment was a reward to the pro-life religious conservatives as he had consistently opposed invitro fertilization, cloning and other medical technologies on the grounds that they rob us of "human dignity." Kass made opposition to all human enhancement technologies, from pharmaceuticals to genetics, the principal agenda of the PCB.

Kass appointed other conservatives to the PCB to further the mission, such as Francis Fukuyama, who published his attack on enhancement technologies in 2002, *Our Posthuman Future.* In October 2003, the Council released its own report critiquing the prospect of human enhancement, *Beyond Therapy: Biotechnology and the Pursuit of Happiness.* In *Rapture,* a history of transhumanism, Brian Alexander says that Fukuyama and Kass are the "Vince Lombardis of the biotech opposition, insisting that suffering was a part of nature. Hurt? Tape it up. Happy pills like Prozac were sucking the manliness out of us. So what if you're depressed? That's life. Kid can't pay attention? Well, some kids are like that. Stop being sissies. Take your imperfections like the true Americans you are and die on time."

Now that the debate about human enhancement and posthumanity has moved from the fringes of cyberculture, science fiction and bioethics to the apex of federal policy debate, the defense is also stirring. Corporate apologists and libertarians have ridiculed all concerns about technology, from the absurd to the deadly serious, and regular defenses of genetic technology appear from those quarters. But bioLuddism has also spurred a spirited defense of genetics and human enhancement from mainstream bioethicists and policy analysts. For instance, in 2000, four leading American bioethicists—Allen Buchanan, Dan Brock, Norman Daniels and Daniel Wikler—published *From Chance to Choice: Genetics and Justice,* which defended parents' rights to use genetic engineering to enhance their kids. In 2002, UCLA philosopher Gregory Stock published the much more widely read *Redesigning Humans: Our Inevitable Genetic Future,* which again defended parental rights to enhance kids with "germinal choice technologies." Now there are essays, books and organizations emerging from many quarters defending human genetic engineering and the project of human enhancement.

The vanguard of this emerging pro-enhancement coalition is "transhumanism." The term transhumanism was popularized in cyberculture by enhancement technology enthusiasts, and there are now dozens of explicitly transhumanist groups and publications around the world. Transhumanists extend the liberal democratic humanist tradition to a defense of our right to control our own bodies and minds, even if our choices make us something other than "human." Transhumanists believe liberal democracy can and must accommodate the "posthumans" that will be created by genetic and cybernetic technologies.

After 400 years of democratic citizenship gradually being extended to include working men, women and all races, transhumanists argue that rights should now be extended to all self-aware minds. In bioethics, this view is known as "personhood theory": Only self-aware persons can have rights. "Persons" don't have to be human, and not all humans are persons. The bioLuddites, on the other hand, advocate "human-racism," the idea that only human beings can be citizens. Just as European racists once denied citizenship to Africans and believed that allowing Africans to have rights would destroy democracy, the human-racists want to deny citizenship and the right to exist to posthumans, intelligent animals and robots.

Beyond broad agreement on individual autonomy and the rights of persons, the differences among the transhumanist-inclined are as profound as the differences among bioLuddites. Some transhumanists are libertarians, closer to the *Wired,* cyber-libertarian, technophile subculture, dismissing risks and arguing for a free market in genetic enhancements and nanotech. Other transhumanists, like Buchanan and colleagues, favor a social welfare state, with public health regulations, national health insurance and efforts to maintain a rough equality among humans even as they upgrade themselves.

The strength of the various flavors on each side will help determine the outcomes of the struggles to come. For instance, in the struggle over embryo stem cell cloning, the bioLuddite side is led by the well-organized but politically unpopular right-to-life movement, while a huge coalition of patient advocacy, provider, scientific and public policy groups are defending stem cell research for therapeutic medical purposes. If the women's health groups opposed to embryonic stem cell research on feminist and safety grounds were more prominent on the bioLuddite side, and the self-interests of biotech firms and medical researchers were more visible in the defense, the outcome might be very different.

This book proposes that the best path toward a transhuman future is the middle way pointed to by Jonathan Glover two decades ago, and by Buchanan and colleagues more recently, a path that addresses bioLuddite concerns about safety and equality and avoids the reckless free-market ideology of some on the transhumanist side. We can embrace the transhuman technologies while proposing democratic ways to manage them and reduce their risks. For the new transhuman era to fully empower people politically and technologically we need a democratic transhumanist movement fighting both for our right to control our bodies with technology, and for the democratic control, regulation and equitable distribution of those technologies.

Section One (Chapters 1 to 5) describes the first steps we are now taking into the transhuman future, and how important both technological and political self-determination have been thus far. Chapter 1 outlines the basic insight that technology and democracy are the principal ways we have of improving the quality of life. Then I explore the four motivations leading us to push the envelope of humanness. Chapter 2 looks at the contributions of democracy and technology to achieving control over our bodies, including less pain, disease and disability. Chapter 3 looks at what we know about living longer. Chapter 4 explores how we get smarter. Chapter 5 asks whether it all will make us happier.

In Section Two (Chapters 6 to 10) I argue that the emerging "biopolitical" polarization between bioLuddites and transhumanists will define twenty-first-century politics. The biopolitical axis joins the existing two axes used in political science, cultural and economic politics. By teasing out these three axes in Chapter 6 we can better understand the flavors of transhumanism and bioLuddism already emerging.

Chapter 7 outlines the central political argument between the bio–Luddites and the transhumanists, the bioLuddites' "human-racism" versus the transhumanists' personhood-based "cyborg citizenship." The implications of these two views are explored in debates about how to deal with the beginning and ending of life, and with animals modified for intelligence, intelligent machines and posthumans. In each case catalytic technologies, such as artificial wombs, the repair of brain injuries with prostheses and the enhancement of animal intelligence, will force us to choose between pre-modern human-racism and the cyborg citizenship implicit in the liberal democratic tradition.

Chapters 8, 9 and 10 describe various biopolitical strains and groups. Chapter 8 reviews how the religious Right has added transhumanism to its list of dangerous social movements, alongside secular humanism and gay rights. Chapter 9 reviews the contributions of deep ecology, feminism, disability rights and anti-capitalism to bioLuddism. Most of the arguments focus on inheritable corrections and enhancements of the human genetic code, but the arguments generally also apply to cybernetic and pharmaceutical enhancements of human abilities.

I discuss the "precautionary principle," which argues that technologies should not be used until their risks are understood, and is often used to argue that humans should not be allowed to enhance themselves until we understand all the consequences. The catch, of course, is that we can never understand all the consequences of any technology.

I also note the argument that enhancing our kids' intelligence or health is some form of totalitarian control of their lives, which has been made by thinkers as diverse as C. S. Lewis and Jürgen Habermas. In response I argue for "procreative beneficence" and "procreative liberty": We have an obligation, both as a society and as parents, to make sure our kids are as healthy, able-bodied and intelligent as possible. At the same time governments have to meet a very high burden of proof before they can justify interfering in parental decision-making about how, when and what kind of children to have, especially in light of the history of coercive eugenics. The best solution then is to make genetic therapies and enhancements as widely available as possible, and leave it up to the general goodwill of parents to act in the best interests of their children.

Chapter 10 examines the emergence of transhumanism. I start by outlining how transhumanism is the synthesis of the ancient drive to achieve healing, longevity and transcendence, previously sought through magic and mysticism, and the ancient drive to explain the world in natural terms and control it with reason. These two strains meet in the Renaissance in liberal humanism, which celebrates the possibility of human self-improvement through science and reason. Then, in the post–World War Two period, the converging influences of bioethics, science fiction, life extension medicine, artificial intelligence and space exploration converge with secular humanism to birth transhumanism as we know it.

I outline how several of the recent flavors of transhumanism were shaped by the cultural and political backgrounds of their exponents. For instance, extropianism rose out of young, male, Californian, libertarian culture with the emergence of the World Wide Web, followed by the broader, more democratic version of transhumanism embodied in the World Transhumanist Association (WTA), which emerged among a more diverse group of European transhumanists.

In Section Three (Chapters 11 through 14) I describe my vision of a "democratic transhumanism." Chapter 11 returns to the principle I illustrate in Section One: People will be happiest when they individually and collectively exercise rational control of the social and natural forces that affect their lives. The promise of technological liberation, however, is best achieved in the context of a social democratic society committed to liberty, equality and solidarity. The present chaos and misery of societies without functioning governments show the absurdity of the libertopian fantasy of freedom from government: We can only be free, prosperous, equal and safe under effective, accountable government.

Much of Chapter 11 is framed as a conversation with bioLuddite left-wingers on the one hand and the libertarian transhumanists on the other. I make the case for why political progressives should embrace human enhancement, and why libertarians sympathetic to human enhancement need to embrace democratic governance. Then I discuss the various constituencies that a democratic transhumanist movement needs to build coalitions with, such as advocates for reproductive and transgender rights, the gay and lesbian and disability communities and advocates for a guaranteed basic income.

In Chapter 12 I sketch out some of the policies suggested by democratic transhumanism. I flesh out the idea of cyborg citizenship, and what the idea says about what we owe to "disabled citizens" such as children, cognitively disabled adults and great apes. I address the limits of biological property and the patenting of human genes. I argue that we need stronger safety testing and regulation of human enhancement technologies, but that these can be handled within the existing regulatory agencies. We do not need new agencies whose sole purpose would be to ban technology on the basis of vague and spurious anxieties.

Once technologies are determined to be safe, we need to figure out which ones should be available in the market, and which ones should be provided universally to all citizens. I suggest the use of the "quality-adjusted life year" (QALY) as a way of thinking about what should be included in a universal health care system. The QALY idea can free health care priority-setting from the faulty therapy-enhancement distinction championed by the bioLuddites.

In Chapter 13 I discuss how we might guard against the development of antisocial personalities with posthuman powers, in ways that still respect cognitive freedom. I call this the Magneto-Xavier problem, which is a reference to the mutant supervillain leader in the comic-book series *X-Men*, Magneto, who decides that mutants are superior to human beings, while Professor Xavier leads the mutants working for human–post-human solidarity.

In Chapter 14 I sum up and present a twelve-point agenda for building a democratic transhumanist movement, including our need for transhumanist organizations and a future-friendly Left, as well as the need to build coalitions with key movements, to join campaigns for closely related rights and social programs and to support transhuman sciences.

Within a couple of decades the issues and ideologies I discuss in this book will seem archaic, since it is impossible for even the most visionary futurist to escape the limitations of time and place. John Haldane, an early-twentieth-century scientist who is one inspiration for a democratic transhumanism, said the future will be "queerer than we can imagine." But in my moments of awe-full reflection on the far future of intelligent life, I think of an image from the Chinese Buddhist Avatamsaka Sutra. The sutra describes the society of enlightened beings as an infinite net, laced with pearls and gems, each enlightened mind a multicolored twinkle that is reflected in every other jewel.

My hope is that, whatever forms of intelligence we give birth to, and however long and strange the trajectory of those embers of mind we fling into the universe, they continue to respect their separateness and diversity, as well as their interconnectedness. Whether they all still believe in *"Liberté, Égalité, Fraternité"* when they are journeying between the stars, I hope they treat one another with the kindness that comes from understanding that other beings are also looking out at the world from finite minds. I believe firmly that the decisions we make in this century, whether to end war, inequality, poverty, disease and unnecessary death, will determine whether we achieve that long-term, inconceivable destiny of intelligence, or flicker out as a failed experiment. Hopefully this book will contribute to a more positive outcome.

TOOLS FOR A BETTER YOU

Better Living Through Science and Democracy

The liberal democratic revolution, centuries old and still going strong, has at its core the idea that people are happiest when they have rational control over their lives. Reason, science and technology provide one kind of control, slowly freeing us from ignorance, toil, pain and disease. Democracy provides the other kinds of control, through civil liberties and electoral participation. Technology and democracy complement one another, ensuring that safe technology is generally accessible and democratically accountable. The convergence of nanotechnology, biotechnology, information technology and cognitive science in the coming decades will give us unimaginable technological mastery of nature and ourselves. That mastery requires radical democratization.

THE NANO-BIO-INFO-COGNO REVOLUTION

In 2002, Jonathan Trent and Andrew McMillan, researchers at the NASA Ames Research Center in California's Silicon Valley, were quietly working to change all our lives. In vats in their lab they cooked up batches of genetically engineered bacteria for use as microscopic engineers. They did this by genetically combining *Escherichia coli* bacteria, which makes people sick in undercooked food, with *Sulfolobus shibatae*, an "extremophile" bacteria discovered growing in the acidic boiling mud of hot springs. The hot springs bacteria produces a protein that self-assembles into an extremely small latticework capable of binding metal into thin wire patterns on silicon. Mixing the *S. shibatae* with *E. coli* made it possible to grow large amounts of the new hybrid. In other words, the hybrid bugs may eventually make it possible to mass-produce extremely small computer chips.

This is one of thousands of lines of research around the world that drives the trend pointed out by Gordon Moore, the co-founder of the computer chip company Intel. "Moore's Law" is the observation that the number of transistors in computer chips and the resulting processing power of computers has doubled every 18 months since the early 1970s. This exponential trend line can actually be traced all the way back to the abacus and slide rule. The shrinking of silicon chip components has allowed more and more transistors to be packed into each chip, but in twenty years or so this exponential doubling will reach the electromechanical limits of silicon. So researchers have been looking for materials to make even smaller chips, materials such as the nanowires excreted by the NASA Ames superbugs.

If these new materials permit the doubling rate of computing power to continue, ordinary computers will quickly become as powerful as the human brain. Computer scientists Hans Moravec and Ray Kurzweil have generously estimated that the human brain is capable of about 100 trillion calculations per second, or 100 teraflops. At the current rate of doubling of computer power, consumer-level electronics will have chips that perform 100 teraflops by the 2020s. But the rate of innovation itself appears to be accelerating, and we might get to human brain-level machines even sooner. IBM has already built a 100 teraflop supercomputer and says it will be making 1 teraflop computer chips for consumer use by 2010.

Just making as many calculations as the brain doesn't generate intelligence, however, since brains and computers have very different architectures. But researchers are also using our growing understanding of neurology to simulate brain structures. Artificial Developments has built CCortex, for instance, a supercomputer that simulates the 20 billion

neurons in the human cortex, including specialized regions of the cor-
pus callosum, amygdala and hippocampus. CCortex also interacts with
simplified versions of the brain stem, thalamus, cerebellum, retina and
auditory processing systems. Even if it takes another decade or two to
make hardware as flexible as neurons, and software as robust and complex as human consciousness, we will create human-level artificial intelligence before the middle of this century.

At the same time, computers will shrink and escape off our desks
into our clothes, our glasses and our built and natural environment. By
2030, people in the industrialized world will live in a sea of cheap, ubiquitous computing extending from their bodies and networking them
with one another and the world. By the middle of the century, the global
network of computing power will be astronomically bigger than the
computing power of all human brains put together, if such a distinction
still makes sense. Even if one is a skeptic about the ability of human beings to create self-aware machine minds, with all that information complexity in the world we can be sure that some very interesting things will
begin to emerge without any conscious human design at all. There will
be ghosts in the machines.

So the research on next-generation computer architecture, at Ames
and elsewhere, is essential for bringing about the exponential increases
in computing power, and all that will bring. But the silicon-spewing bugs
at Ames also represent a second very important trend—the convergence
and integration of biotechnology, molecular manufacturing and computing. As the knowledge and progress of each of these different scientific fields grow exponentially, they will frothily interact and build off of
one another.

For instance, thanks to computing advances, the speed at which all
the information on human genes can be read into a computer has been
increasing exponentially, and its cost shrinking exponentially. In 1965, it
cost more than $1,000 to sequence one of the 3 billion base pairs in the
genome. By 1991, up to 10,000 bases could be read in a single day for
about a dollar a base. It took the Human Genome Project ten years and

$3 billion dollars to sequence the first complete genome, which it finished years ahead of schedule in 2000. The costs of gene sequencing are now being cut in half every two years or so.

Now, just a few years after the Human Genome Project finished, firms are working on machines that will rapidly and inexpensively read out all of a patient's genetic information, permitting the diagnosis of genetic diseases and risks and the prescribing of drugs uniquely tailored to our genes. "Personal genomics" is benefiting from new machines that can unravel chromosomes and quickly read them one base pair at a time, and from micro-labs on computer chips that use millions of channels and chambers to sequence genes and read out their structure. In the next decade or two, cheap, instant genetic identification and analysis will also revolutionize criminal justice by making the genetic analysis of crime scene data routine.

Nanotechnology and the miniaturization of robotics are a third area in which exponential trends suggest that things will be getting very small, very fast and very cheap in the very near future. Current nanotechnology research is focused on building simple nanomaterials and nanodevices, all at the scale of 10 to 100,000 times smaller than the diameter of a human hair. Each node in the computer chips that the Ames hybrid bugs produce is about 20 nanometers wide, a quarter the size of the 90-nanometer nodes used in 2003 computer chips. In 2003, nanotechnology firms announced they had developed nanorobotic arms that can be operated remotely to control objects as small as 100 nanometers. Extrapolating our rapid progress at shrinking computer chips and robotics, at some point in the coming decades, we will be able to build robots the size of viruses with onboard computing, telecommunications and nanoscale manipulators.

When we program these nanorobots to use available materials to build more copies of themselves, we will have achieved the holy grail of nanotechnology: the "molecular assembler." Vats of nanorobots will be grown and injected to roam around in our bodies inserting genes, eliminating cancers and interacting with our brains. We will put other molecular assemblers in a bucket and program them to make things, from the molecules and atoms up. Since viruses, cells and genes are basically successful organic nanorobots, and since biology, computation and nanofabrication are rapidly merging, we may have nanorobots even sooner than the engineers expect. Computer models are being created of the entire genomes of viruses and microorganisms and every one of their functions, allowing computer-aided design to reengineer them into nanorobots.

Trying to predict the progress of these evolving and cross-fertilizing fields requires a good bit of humility. But even the most sober of scientists are beginning to get very excited. In 2002, the U.S. National Science Foundation (NSF) convened a series of workshops and commissioned a series of papers on the consequences of the convergence of nanotechnology, biotechnology, information technology and cognitive science (NBIC) for "improving human performance." The project drew together more than a hundred experts on science and technology from government, academia and business, and their enormous report, available online, has become an instant landmark in near-future technological speculation. The report includes detailed plans for nanotechnology to be intimately woven into the body and brain in the very near future. The NBIC report concludes: "With proper attention to ethical issues and societal needs, converging technologies could achieve a tremendous improvement in human abilities, societal outcomes, the nation's productivity, and the quality of life."

This book is about how to bring together those "ethical issues" and "societal needs" with the technologies that will so radically change our day-to-day existence. Rather than focus on some timeline for all of these technological wonders, a timeline that would be out of date by the time you read these words, I want instead to focus on the constants, the things unlikely to change in the coming century: the human needs and desires these technologies will be asked to serve.

The four motivations or applications I focus on in this section are our desire to control the body, to live longer, to be smarter and be happier.

The NBIC technologies will definitely also change how we work, how we travel, how we communicate, how we worship and how we cook. But the most fundamental changes in our lived experience will come from their impacts on our bodies and brains. They will also create fundamental challenges to the politics of the twenty-first century and raise basic questions about citizenship, freedom and equality.

Will these technologies be designed to serve human needs and not just corporate and military ends? Will these technologies be suppressed and stunted by irrational fears and religious dogmas, or affordable only by the wealthy, or sold to us with inadequate testing for safety and efficacy? Will they threaten the environment, or even threaten our souls?

DEMOCRACY AND TECHNOLOGY

The central propositions of this book are that people are generally happier when they have more control over their own lives, and that technology and democracy are the two key ways by which we can exert more control over our lives. In the last century, agricultural and industrial technology made it possible for most people in the developed world to leave the farm and consider dozens of possible jobs, instead of the three or four that would have been open to their forebears. But the freedom from agricultural labor was also made possible by the abolition of slavery, universal education and rural electrification, all the fruits of democracy. Industrial employment would have been little improvement over sharecropping without the labor movement, workplace safety regulation and unemployment insurance, also products of democracy. Technology has put entire libraries of information at our fingertips, and public education, public libraries and freedom of the press have made that information accessible. Contraception and abortion made family planning and sex for pleasure possible, but we first had to win the right to use them.

Unfortunately, many people around the world are still not permitted to pursue education, access the Internet or get an abortion, and are too poor to exercise these freedoms in any case. Most of the world's poor are still struggling to build the high-tech economy and democracy that evolved together in the developed nations over centuries. Industrial development made possible the growth of the working and middle classes, which in turn demanded suffrage and rights. The spread of scientific worldviews weakened religious and political authoritarianism, and encouraged pluralistic tolerance.

The pluralistic democracies, in turn, encouraged the rapid development of science and technology. Although the Soviet and Nazi governments invested heavily in scientific research and industrial development, they also crippled the sciences with state interference. Today the government of China is straining to open their economy to the global flows of information and people necessary for technological development while maintaining its authoritarian political control. Authoritarian regimes try to clamp down on their citizens' use of condoms and computers that might threaten patriarchy or the state, whereas democracies focus on making sure that the technologies are safe and widely accessible.

The idea that control over the conditions of life is the path to happiness might seem a peculiar proposition for a Buddhist like me. Isn't all life suffering, and spiritual peace found in just accepting the way things are? Perhaps all this technology is a distraction and we should accept the "human condition" as it is. If we really take that approach, why do anything? Why try to improve the world or our own lives with technology, politics or any effort at all? I think the real insight of Buddhism, and all the other religious traditions, is perfectly expressed in the Serenity Prayer written by theologian Reinhold Niebuhr:

> *God, give us grace to accept with serenity*
> *the things that cannot be changed,*
> *courage to change the things*
> *which should be changed,*
> *and the wisdom to distinguish*
> *the one from the other.*

In other words, diet if you need to, but don't make yourself miserable if you aren't very successful. Work for your favorite candidate, but don't go into a deep funk when they lose. No matter how much progress we make there will always be something that is beyond our grasp, something that we can't control that we will have to find the serenity to accept. In the meantime, there is a lot of overt suffering in the world that we can control. We can stop violence, war and torture, and use medical technology to repair bodies and brains. We can clean up the toxins that cause cancer, and use the new technologies to cure the cancers that they cause. We can treat clinical depression with effective drugs, and create a society that causes people as little suffering as possible. Using technology to live longer, smarter and happier lives is not a substitute for spiritual wisdom, but it isn't a distraction from wisdom either. In fact, ensuring the world has access to technologies that reduce suffering may be a spiritual obligation.

In the next four chapters I discuss how both technology and democracy have contributed, and will contribute in the coming decades, to our controlling the body, living longer and getting smarter and happier.

Controlling the Body

In democratic societies we try to give each other as much control over our own bodies and minds as possible. Now technologies promise to make it possible to fully reconfigure our bodies and minds. Will we make it possible for everyone to use these technologies to achieve their fullest capabilities?

In *On Liberty,* John Stuart Mill wrote: "The only part of the conduct of any one, for which he is amenable to society, is that which concerns others. In the part which merely concerns himself, his independence is, of right, absolute. Over himself, over his own body and mind, the individual is sovereign." In the Western democracies we have more or less accepted Mill's argument. We have institutionalized the idea that doctors must secure "informed consent" before tests and procedures. Women and children are protected by law from violence and abuse to an unprecedented degree. People may move and emigrate freely, and generally speak their minds. There are few remaining laws against "victimless crimes," with the painful exception of the War on Drugs and criminalization of sex work, and even there most democratic countries are liberalizing their drug and sex work laws.

Soon, unlimited technological control over the human body and mind will be possible and the battle over how much of that control individuals should be allowed to exercise will dwarf the fights over drugs and prostitution. But the first beneficiaries of those technologies will be the sick and disabled, for there is little controversy that they should be able to use technology to more fully control their own lives.

PREVENTING AND FIXING DISABILITIES

The percentage of the population that is disabled continues to shrink in the developed world. The trend began more than a century ago and is directly related to the spread of democracy. The labor movement and occupational safety and health laws reduced workplace injuries. The consumers movement campaigned, and federal agencies mandated seat belts and air bags for cars, and screening for dangerous foods and medicines. Public health systems stopped the spread of infectious diseases so fewer people were left blind, deaf, crippled and dead in each decade. Hospitals, built with and maintained by public monies, applied advanced trauma techniques to save lives and limbs in their emergency rooms.

The percentage of the population born disabled has also been shrinking. Improvements in prenatal and neonatal care have shrunk the percentage of babies born with preventable disabilities. Mothers have increasing access to prenatal diagnosis and abortion and can choose not to bring disabled fetuses to term. The prognosis and treatment options for babies born with problems have also improved rapidly with the growth of neonatal intensive care units and pediatric medicine.

The decline of disability has further accelerated in the last two decades as seniors have benefited from rapidly improving medical and nursing care, paid for by their governments. Even though the number of American senior citizens increased by more than 30% between 1982 and 1999, there were fewer disabled seniors in 1999 than there were in 1982. Seniors' disabilities today are also less likely to be chronic. In 1982, more than 25% of American seniors over 65 suffered chronic disabilities, but that had shrunk to less than 20% in 1999.

This decline of disability not only improves the quality of life of seniors and their caregivers, but it reduces social costs on society. The declining rate of senior disability is especially good news for the soon-to-retire Baby

Boomers, who demographers keep predicting will break the social safety net. The U.S. National Center for Health Statistics reported in 2003 that the average senior citizen with no activity limitations cost about $4,600 in health care, while the moderately disabled senior cost about $8,500 per year, and nursing home residents cost about $45,000 per year. The declining burden of active, healthy seniors on the health care and nursing system means that those who live more able-bodied years cost no more overall than those who live fewer sicker years.

Unfortunately, the percentage of the population that is disabled has declined only in the developed world. The rates of disability in the developing world are much higher since they are not affluent enough to make use of Western medical technology, and their governments aren't powerful and accountable enough to enforce Western levels of worker safety and consumer protection. In the absence of strong public health systems, hundreds of millions of people in the developing world suffer from curable and unnecessary forms of blindness, deafness and physical disability.

HAVING HEALTHY, ABLE-BODIED CHILDREN

An estimated 7.6 million children are born each year around the world with congenital abnormalities, births that could have been prevented if parents had access to prenatal testing and abortion. Many of these children die, but those who don't are often disabled. For instance, the first prenatal screens for Down syndrome, the most common form of retardation, became available in 1968, and today about 65% of American births are screened for Downs. This has reduced the rate of newborns with Downs by 50%. But the prenatal screening methods, alpha-fetoprotein and amniocentesis, and the subsequent abortions are all expensive, and in many countries illegal. As a result only 7% of births are tested for Downs worldwide, and even in the United States poor and uninsured women who need prenatal screens are less likely to receive them.

Fortunately the painful decision to abort a disabled fetus will soon be made unnecessary by our genetic and reproductive technologies.

Already we are able to test embryos conceived with invitro fertilization, a technique called pre-implantation genetic diagnosis (PGD), allowing parents to choose only healthy embryos for implantation. Thousands of parents have used PGD now in the U.S. and Europe. Unfortunately invitro fertilization and PGD are still so difficult, uncertain and expensive that, for the next decade or two, only the affluent or lucky few who use them as part of infertility treatments will benefit from PGD. But as the cost of invitro fertilization drops, and its success rate climbs, PGD is likely to become more common.

PGD, genetic screening and prenatal testing are just a few of many "germinal choice" technologies parents will be able to use in the coming decades. Parents can already use donor sperm or eggs if one or the other partner can't contribute genetically to making a child. Germinal choice will also soon include direct genetic engineering of the parents' sperm or eggs before they reproduce and of the early embryo itself.

Genetic therapies that change parents' sperm or eggs, or the early embryo, also change the reproductive cells of the child, their "germline." So these are known as "germline therapies." Germline therapy or enhancement is very controversial since critics believe we can never understand the genome well enough to avoid unintended consequences for future generations. Critics of germline modification want gene therapies to be restricted to "somatic" modification, that is, fixing genetic diseases only in nonreproductive cells by, for instance, infecting them with viruses carrying new genes. But somatic therapies fix only the tissues they can reach, so instead of removing cystic fibrosis from all of someone's genes we would be required to regularly tweak their lung cells, wait until their immune system killed all the corrected cells, and then treat them again. Germline therapies are comprehensive and permanent. People modified at the germline level would never need repeat treatments and could pass their cleaned-up and upgraded genes to their children.

A shortcut around some of the safety and ethical objections to germline therapies comes from "artificial chromosomes," a method of human germline engineering being promoted by philosopher Greg Stock. Researchers at the Canadian firm Chromos Molecular Systems have created complete, free-standing chromosomes inscribed with genetic instructions to make specific proteins and simply added them to the nucleus of cells alongside the other chromosomes. This eliminates the risk that inserting genes at random into existing chromosomes will screw up something unforeseen in our DNA. The artificial chromosomes can be designed with more numerous, longer and more complex genetic instructions than are possible with existing "insertional gene vectors." The artificial chromosomes can also be fitted with chemical switches so that parents, or the child herself when she is of age, can turn the new genes on and off.

For instance, neurogeneticist Kathryn North at the University of Sydney in Australia has found that good sprinters have one copy of the alpha-actinin-3 gene, which makes proteins for "fast-twitch" muscle fibers, and that great sprinters have two copies. Long-distance runners usually have the alpha-actinin-2 gene, good for "slow-twitch" muscles fibers, and great endurance runners have two copies. Why not add a couple copies of each on an artificial chromosome with on/off switches? Ideally the kids will be able to sprint and run long distances, and if it turns out the genes turn them into the Hulk, we could just turn them off.

While I believe the technical and ethical obstacles of human genetic engineering will quickly be surmounted, it is certainly true that progress with gene therapy has been slower than was expected in the 1990s and has suffered many setbacks. Many biotech firms testing gene therapies rocketed on a wave of optimism, and then crashed and burned with the rest of the 1990s' bubble economy. Many clinical trials of gene therapies disappointed when genes didn't make it into cells, only to be quickly expelled by the body, sometimes triggering cancers or life-threatening allergic reactions. In 1999, 18-year-old Jesse Gelsinger at the University of Pennsylvania died when his immune system rejected a gene therapy trial. In January 2003, two of gene therapy's success stories, French boys cured of a rare immune disorder, contracted a rare leukemia, leading to the temporary suspension of more than thirty American gene therapy trials. After the startling breakthroughs in cloning in the 1990s, further progress has been slowed as many mammal clones have turned out to suffer from gross genetic deformities.

These setbacks have spurred a quite legitimate regulatory crackdown worldwide. The National Institutes of Health has tightened requirements on the reporting of adverse events by gene therapy researchers, and the Food and Drug Administration has asserted its right to forbid germline genetic modification and human cloning experiments that have not been deemed safe. More extreme voices have begun pressing for national and global bans on inheritable germline modifications and therapeutic stem cell cloning. But even with these research and political setbacks, widespread somatic and germline engineering, for both correcting genetic defects and adding enhancements, is virtually certain to be available within a decade or two to those who can afford it. The only real questions are how sure we will be of the techniques' safety and efficacy, and how accessible they will be.

Even if we were to forbid intentional, inheritable genetic modification of sperm, eggs and embryos, the issue is likely to be made morally and technically moot by effective somatic therapies. An effective somatic gene therapy, injected into seminal vesicles, ovaries or uteruses, would be able to change eggs and the cells that generate sperm. Even if patients do not intentionally attempt to modify their own germinal tissues it will be difficult to keep effective somatic therapies away from the reproductive cells; the whole point is to get the gene inserters to survive and change lots of cells. Once someone accidentally or intentionally tweaks their own sperm or eggs while fixing their own cystic fibrosis or cancer genes, I doubt we in the democratic West would pass laws forbidding them from having children. So, if we agree that people have a right to control and change their own reproductive tissues, then germline modification, both therapeutic and enhancing, is more or less inevitable.

BEYOND DISABILITIES TO CYBORG SUPERPOWERS

What if the most pessimistic naysayers turn out to be correct, however, and the genome's complexity makes safe gene therapy impossible for fifty or a hundred more years? Without gene therapy, some disabilities will still be eliminated, and we will still all have the option of acquiring superhuman abilities, thanks to cybernetics and nanotechnology. Today 25,000 deaf people worldwide have implanted, computerized hearing aids—"cochlear implants"—sending electrical impulses directly to their auditory nerves. The rapidly improving cochlear implant is already making it possible for deaf kids to be mainstreamed back into hearing schools. Current models spiral around the cochlea pressing dozens of electrodes against auditory hairs.

Owing to interference between the electrodes, most patients have only four to seven functional channels or frequencies and a very poor auditory signal. As progress in nanomaterial science shrinks the electrodes, more electrodes can be put closer to or directly into the nerve, stimulating a narrower band of frequencies and creating a more precise sound palette. Smaller electrodes will also require less energy, permitting a completely implantable device, without the microphone pickup in the pocket. Eventually the capabilities of the cochlear implant will reach and surpass the auditory fidelity of the unaided inner ear. Hopefully, as cochlear implants improve, they will become cheap enough to be accessible to the millions of deaf worldwide. But even today in the United States, there are still approximately seven million people who could benefit from simple hearing aids who cannot afford them.

The new electronic vision systems for the blind are even more expensive. The Dobelle Vision System, which captures video on digital cameras and transmits to electrodes inserted in the visual cortex, has allowed its first blind recipients to see well enough to drive. Installing and maintaining the Dobelle system currently costs more than $100,000, including surgery, follow-up, cameras, microcomputer and warranties. But the costs will eventually drop, the size of the equipment will shrink until it all fits in the eye and the acuity of the vision will eventually surpass ordinary vision. Other teams are working on solar-powered systems and electrodes that can stimulate individual neurons. Eventually these systems, like Geordi LaForge's wraparound visor in *Star Trek: Next Generation,* will permit the blind to see ultraviolet and infrared light invisible to the biological eye—at least for people lucky enough to live in societies that allow them to have these super-eyes, and rich enough to make them available.

The repair of spinal cord injuries, which afflict 250,000 Americans, and tens of millions of people around the world, is also very close, as the quadriplegic activist Christopher Reeve has made clear through tireless advocacy. Most of the research on spinal cord repair has focused on the use of stem cells to regrow severed spinal cord connections. Stem cells were first isolated and cultured from adults and embryos in the 1990s. Stem cells are baby cells, not yet differentiated into the kind of tissue they will end up becoming. In embryos these cells are very "plastic," and can become virtually anything. The stem cells generated in adult bodies are more limited, but can still be adapted to repair damaged tissues.

Researchers have shown that both embryonic and adult stem cells can be injected into Alzheimer's-ravaged brains, severed spinal cords and diseased livers, and the cells migrate to the damaged area and become the kind of cell that is needed there. Research is progressing furiously around the world to develop treatments using cultured stem cells. That is, everywhere but in the United States, where the Bush administration, under pressure from the powerful anti-abortion lobby, has forbidden funding of research that uses new lines of embryonic cells. Nonetheless, heavy investments in China, Britain and elsewhere ensure that embryonic stem cell research will continue to progress.

PURITAN RESISTANCE TO BODY CONTROL

Making the deaf hear, blind see and lame walk are relatively uncontroversial applications of emerging technologies, even if some of the re-

search methods, like modifying genomes or using embryos, are controversial. But new technologies will also permit people to reshape their
bodies to fit their personal aesthetics, lifestyles and whims, and this of-
fends the deep Puritan strain that runs through the emerging bioLuddite
movement. The growing battle over the epidemic of life-threatening
obesity is one place where we can clearly see the bioLuddites' Puritan
anxiety that control over the body will facilitate sin.

For instance, Pat Mooney, head of the anti-technology lobby "ETC,"
attacked anti-obesity drugs in 2002 in *WorldWatch* magazine:

> The pharmaceutical industry is hard at work developing drugs that
> allow people to eat gluttonously without getting fat. . . . Of course the
> logical solution is to eat less and exercise more. But there is a multi-
> million dollar market waiting for any pharmaceutical company that
> can . . . let people stuff their faces without losing their figures . . . the
> real goal is a magic elixir that turns indulgence into a virtue.

One expects him to then warn about the dangers of masturbation, and
the importance of an early bedtime.

Greg Critser's bestseller *Fatland* argues that the obesity epidemic was
generated by new food technologies, such as corn syrup and palm oil,
aggressively promoted by an underregulated, profit-driven food industry.
The fat and sugar contents of food increased, as did the size of portions,
while federal bureaucracies, undermined by food industry money and
lobbyists, made no effort to encourage healthier food choices. At the
same time industrial society allowed us to become less active, and encouraged more sedentary entertainments. Critser recommends restricting advertising that targets kids, removing fast food and soda from the
schools, expanding physical education requirements in schools, and in-
stituting a fat tax on unhealthy foods.

All good ideas, and corporations are certainly eager to make profits
by first helping us kill ourselves and then by fixing us back up. But the
basic cause of obesity is that we have bodies designed to spend hours

walking around the savanna every day, and brains that find easy access to fats, sugars and carbohydrates irresistible. Only safe and cheap genetic and pharmaceutical therapies can successfully stop the deadly world-wide rise of obesity.

Fortunately those therapies are coming soon, and I've certainly been anxiously anticipating their arrival. More than sixty pharmaceutical treatments to alter metabolism or reduce appetite are being developed, based on more than 130 genes that have been discovered to regulate weight in humans. The gene or drug tweaks that keep us slim will likely be much simpler, modifying just one of those chemical pathways. Researchers at the University of Wisconsin–Madison have made mice that can stay slim on high-fat, high-calorie diets by snipping out just one gene, SCD-1, which codes for just one enzyme, SCD, which regulates insulin sensitivity. Research on a database of all Icelanders' genes has turned up one gene that determines whether an Icelander is slim or fat.

Once we have safe, effective fat pills or gene tweaks, the next question will be how many of us will be able to use them. Even if they have few risks, the Puritans will push to make them available only by prescription and only to the morbidly obese. They may also be too expensive for the poor and uninsured. Access to weight-control technology for the poor will be an especially acute social justice issue since the poor are disproportionately fat, making them sicker, shortening their lives and keeping them poor because of severe social stigma and employment discrimination against the obese. The poor are also the least able to purchase diet foods, exercise equipment and over-the-counter weight-loss drugs. Sure, poor people can cheaply eat lots of salad and run around the block, but it's a lot easier to live the simple life if you are rich.

The image of sober moderation is the person who goes to bed early to get their eight hours of sleep. But I have always resented sleep as a weakness, robbing me of a third of my life, and like most Americans I am chronically sleep-deprived. In the past I would have had to rely solely on caffeine or amphetamines. Now modafinil, a new treatment for narcolepsy sold as Provigil, permits people to perform at a peak of attentiveness with just a couple hours of sleep for days at a time, without the side effects of an amphetamine such as twitchiness and mood elevation. While the Puritan bioLuddites fret that workers will be coerced to take alertness drugs by unscrupulous corporations, wouldn't you rather your pilot or that trucker behind you on the road had access to a safe, effective alertness drug?

Puritans are also suspicious of efforts to control gender, sex and reproduction, since they just enable us to fulfill unnatural desires and evade our preordained destinies as baby makers. Today we can control when and whether men get erections, whether sex leads to conception and whether to menstruate. We can plan childbirth from invitro fertilization to cesarean section, and postmenopausal women can bear their daughters' children. Soon baby-making won't need fathers, and artificial wombs will make birth mothers optional. Gay couples will have biological children genetically related to both parents.

People who are unhappy with their birth gender can take hormones, get breasts implanted or removed, have hair permanently removed from their face, and have their vocal cords surgically adjusted. Penises can be reconstructed into sensitive vaginas, and functional penile tissue has been grown in the lab for building new penises. In a couple decades transsexuals will be able to have new, fully functional genitalia cloned, grown and transplanted. Somatic gene therapy will make it possible to selectively turn on or off sex-expressing genes, making hormone treatments unnecessary.

Every step of the way we moderns have had to fight, and will continue to have to fight, the Puritan moralizers for our rights to control our genitals, wombs and baby-making. *Roe v. Wade* established the right to terminate a pregnancy in the United States, and in 2003, the U.S. Supreme Court finally said sex between consenting adults was constitutionally protected. But the right to use contraceptives and abortion are still under attack, and the technologies necessary for reproductive rights are either banned or unaffordable in many parts of the world. Even where transsexuals have the right to change genders, they are often underemployed, victims of violence, and find it difficult to afford their sex-reassignment treatments. Technology makes possible the control of our sexuality, gender and reproduction, but it is an ongoing political fight to ensure that society allows us to exercise those rights, and makes available the resources necessary to exercise them.

At the same time, all of these technologies have risks, and are being developed and promoted by for-profit firms with weak incentives to research and acknowledge risks. A 1993 study by the Office of Technology Assessment (an agency destroyed by the Republicans after they took control of Congress in 1994) found that more than half of the novel drugs brought to market between 1981 and 1988 had been withdrawn because of unforeseen consequences. For consumers to be able to make fully informed choices, and exercise real control over their bodies, we need laws that require these technologies to pass public testing and robust regulation for safety and effectiveness. I sympathize with the bioLuddites' frustration that every regulatory agency has been infiltrated and crippled by the very industries they are meant to regulate, especially under Bush II. But the rational and progressive answer has to be to reclaim the regulatory agencies, not to forbid the use of technology.

3

Living Longer

Life expectancy has dramatically increased in the last century as the developed countries built public health systems. In the twenty-first century we will achieve radical life extension, with indefinite life spans. We need to ensure that everyone has access to these treatments.

THE CONTRIBUTION OF EQUALITY AND PUBLIC HEALTH

In 2002, demographers announced that the average American could expect to live 77 years. That's 30 years longer than the 47 years that Americans could expect in 1900. Unfortunately those numbers hide the fact that life expectancy for American seniors actually changed little in the twentieth century. The real beneficiaries of this dramatic improvement were American infants and children. During the last century, prosperous democratic societies built public health systems and invested in education, clean drinking water, sewer systems, general hygiene and vaccinations. Through public health investments we eliminated the high rate of death from infectious diseases that once killed so many kids.

Toward the end of the twentieth century, public health investments in anti-smoking campaigns, exercise programs and other preventive medicine efforts have also begun to extend life expectancy at the end of life. Not surprisingly the countries with the longest life expectancies are those with the strongest public health systems—social democratic northern Europe and Canada, and relatively egalitarian Japan. People with strong public health systems don't just live longer but are also less disabled. According to the World Health Organization, the average citizen of the United States, which has 40 million uninsured and a weak public health system, can expect two to three *fewer* years of healthy, disability-free life than citizens of northern Europe.

Longevity is associated not only with strong democratic states and public health systems, but also with social equality. Poor people die more often than non-poor people when we compare the developing world to the developed, and the poor within the developed world to the affluent. In all the industrialized countries, which all have universal health care except the United States, the poor die more rapidly, and the degree of a country's equality is correlated with its citizens' longevity. The amount of inequality in each American state is correlated with its rates of infant mortality, with the death rates for all age groups, with the incidence of heart disease and cancer, and with homicide, violent crime and incarceration.

One explanation for the relationship of equality and life expectancy is provided by British studies of a large group of civil servants working in the Whitehall area of London, home of many government ministries. After twenty-five years the researchers discovered that the higher the position the civil servants attained in the bureaucratic hierarchy, the less likely they were to die from all causes, including heart disease and non-smoking-related cancers. None of the men were poor and none were rich, and their work conditions were all pretty similar since they worked in offices in London. All the men, like all Britons, had access to the same National Health Service. Yet the men at the bottom of the work hierarchy, who had less control over their work, died at twice the rate of those at the top.

Even for the affluent, powerlessness is a killer. Poor and powerless people are more likely to feel their lives are driven by chance, by nameless forces and unaccountable elites, increasing their stress and risk-taking. People with a sense of empowerment in their lives are more likely to invest in those lives by pursuing education and making better choices. Having some degree of social and economic equality, some control over one's work and participating in the democratic process all appear to be important to health and long life.

LIVING LONGER ON DRUGS

But even the wealthiest, most empowered person, living in Sweden or Japan, who doesn't smoke or drink, who exercises regularly and eats right, is unlikely to live past 100. In 2003, Majid Ezzati of Harvard University demonstrated that if the world could conquer the top twenty contributors to early mortality—including dietary deficiencies, malnutrition, high blood pressure, high cholesterol, obesity, lack of exercise, unsafe sex, lack of contraception, the use of tobacco, alcohol, and illegal drugs, poor sanitation and dirty water, poor household ventilation, lead exposure, occupational risk factors and unsafe injections—the worldwide gain in life expectancy would be . . . 9 years. Getting life expectancy beyond 100 years will require more than improved public health. Whether the "natural" life expectancy under optimal conditions is 80 or 100 years, the human body is certainly programmed to start slowly falling apart after about the age of 20.

Fortunately we are making dramatic strides in extending life through medicine. While improvements in the quality of life and investments in public health drove the increase in life expectancy in the twentieth century, medical technologies that retard and reverse the aging process will drive radical extensions of life expectancy in the twenty-first century.

This is good news for Americans since we don't do as well at social equality and public health but (at least until the Bush II administration) have excelled at innovating new medical technology. If you can make it to 70 in the United States you have a better chance of living to 80 than anywhere else in the world. Even without the radical life extension medicine around the corner, we have been making slow and steady progress in treating the principal causes of mortality in the industrialized world, namely, heart disease, cancer and stroke. Antihypertensives and cholesterol-lowering drugs have added some years to the lives of those with a high risk of dying from heart disease and stroke. The survival rate of cancers has been increasing, and remarkable new anti-cancer treatments are being developed. We are now able to treat cancer by introducing viruses designed to attack cancer cells and by teaching the patient's immune system to recognize and kill cancerous cells. Each of the genetic mutations that create and sustain tumors, such as mutations that cause rapid cell growth, build a tumor's blood supply or send cells out to spread through the body, is being identified, and specific chemical and genetic therapies developed to counteract them.

But demographers point out that even if these top three causes of mortality—cancer, heart disease and stroke—were completely eliminated, life expectancy would increase less than 15 years. Those who survive these three killers die of other aging-related diseases. As 53 leading aging researchers recently concluded in a review of longevity research published in *Scientific American:* "To exceed this limit, the underlying processes of aging that increase vulnerability to all the common causes of death will have to be modified."

Many lines of research now under way promise to radically slow the aging process or even reverse it. So far, the best documented way to extend the life span of animals is to reduce the number of calories they consume by about a third. This makes sense for evolutionary reasons; if resources are scarce, creatures can hang on long enough to reproduce later. Caloric restriction slows the aging process in tissues throughout the body, and reduces the incidence of many aging-related diseases, including cancer. Calorie-restricted rhesus monkeys have lower cholesterol and less heart disease, diabetes and hypertension. Caloric restriction is not a very practical life extension method, however, since it is even more restrictive than the diets most people have so little success with. So researchers such as Barbara Hanson at the University of Maryland, Cynthia Kenyon at the University of California at San Francisco and Blanka Rogina at the University of Connecticut, are identifying the genes and proteins involved when caloric restriction signals the body to go into starvation mode. This research will soon give us the benefits of caloric restriction in a pill.

Once we have effective methods of changing adults' genes, these calorie restriction mechanisms are potential targets for gene therapies for life extension, and there are probably many more targets to be found. Researchers with the New England Centenarian project and the company Centagenetix are studying 444 families with a family member who lived to 100 or older to identify other genes and chemical processes that could be tweaked to extend life. They say they have found a region of chromosome 4 where a lot of their centenarians appear to be different, and drug firms are rushing to devise treatments that turn on the longevity processes coded for in those genes. Cloned and cultured stem cells also are a potential pathway for replacing worn-out organs and tissues.

Biogerontologist Aubrey de Grey at the University of Cambridge is the founder of the Methuselah Mouse prize awarded each year to the team that engineers the longest-lived mouse. He believes that there is a very strong chance that the convergence of seven biotechnologies will allow us to reverse the effect of aging and achieve indefinite life spans, or "engineered negligible senescence," by the middle of this century. Those technologies are:

1. Stem cell therapy to restore the number of cells in tissues that lose cells with age (such as heart and some areas of the brain).
2. Gene therapy plus stem cell therapy, to delete our biological clock (telomere elongation) genes.
3. Gene therapy to fix the mutation rate in our mitochondrial genes, which contributes to aging.
4. Gene therapy to introduce bacterial or fungal genes that can break down things that we can't such as cholesterol and the proteins that accumulate in the brain and cause neurological deterioration.
5. Therapy to teach our immune system to destroy bad, old cells.
6. Training the immune system to, or introducing drugs to, eliminate the junk (amyloid) that accumulates outside cells.
7. Drugs to break down the sugars that gum up the connections between cells (collagen, elastin, etc.).

There is continual progress in all of these fields, and with an infusion of about $1 billion in targeted funding de Grey believes they could all be perfected in about twenty years.

SEND IN THE BOTS

But what of the convergence of technologies I discussed in the first chapter? Nanotechnology will likely have a huge impact on both the treatment of disease and the eventual arresting of the aging process. Artificial intelligence and robotics will facilitate the biomedical research and protein modeling necessary to develop new therapies, and advances in cognitive science will illuminate the best ways to protect our brains from aging. In the National Science Foundation's NBIC report, Scottish bioengineer Patricia Connolly writes on "Nanobiotechnology and Life Extension," and predicts the advances and their life-extending effects shown in Table 3.1.

Nanomaterials are already creating radical new treatments. For instance, researchers in Berlin have coated nano-bullets of iron oxide with sugar so that they get absorbed into hungry, fast-growing tumors, and then excite the particles with magnets to heat up and kill the tumor cells.

But self-replicating, computerized nanorobotics will offer the most radical life extension. Robert Freitas, the author of *Nanomedicine*, has designed "respirocytes" or nanorobotic blood cells. A couple cups of microscopic respirocytes, each with a small onboard computer to regulate absorption and release, would carry ten times as much oxygen as a normal red blood cell. Oxygen-laden respirocytes in the brain would mean that a person having a heart attack could notice their heart had stopped beating, call an ambulance and still be conscious by the time they were being rolled into the ER. Nanorobot red blood cells could then be supplemented by Freitas's proposed "microbiovores," a nanotech complement for the scavenger and immune system cells, to protect the body from viruses, bacteria and cancerous cells.

TABLE 3.1   Some Potential Gains in Life Extension from NBIC Convergence

Level of Intervention	Key Advance	Timescale	Life Extension
Human	Noninvasive diagnostics	5–10 years	Lifesaving for some conditions
	Cognitive assist devices	15–20 years	Higher quality of life for several years
	Targeted cancer therapies	5–10 years	Reduction in cancer deaths by up to 30%
Organ	Artificial heart	0–5 years	2–3 years awaiting transplant
	Neural stimulation or cell function replacement	5–20 years	10–20 years extra if successful for neurodegenerative patients
Cell	Improved cell-material interactions	0–15 years	Lowering of death rates on invasive surgery by 10% and extending life of surgical implants to patient's lifetime
	Genetic therapies	30 years	Gains in the fight against cancer and hereditary diseases
	Stem cells	5–10 years	Tissue / brain repair Life extension of 10–20 years
Molecule	Localized drug delivery	0–10 years	Extending life through efficient drug targeting
	Genetic interventions	0–30 years	Life extension by targeting cell changes and aging in the fight against disease

I've only touched the surface of life extension medicine. But the many lines of converging sciences and technologies and the rapidly escalating pace of medical knowledge suggest that indefinite life extension will be possible in this century. Since anti-aging and life extension medicine will likely have added at least 50 years to life expectancy by the 2050s, the average child today can expect to live well into the 2100s, when the convergence of cognitive science, nanomedicine and artificial intelligence will allow consciousness to be backed up and sustained in forms far more durable than the human body.

For those who aren't children, and are willing to take a gamble, there is the option of "cryonic suspension," that is, getting frozen. The cryonics movement began in 1964 with the publication of Robert Ettinger's *The Prospect of Immortality*. Ettinger argued that even though the body's tissues suffer considerable damage when frozen in liquid nitrogen, eventually some future technology would make it possible to fix that damage and resurrect the "corpsicle." Cryonics got a big boost in 1986 with the publication of Eric Drexler's nanotechnology manifesto *The Engines of Creation: The Coming Era of Nanotechnology*. Drexler, who is signed up for cryonic suspension, argued convincingly that eventually we will be able to build nanorobots able to travel through a frozen brain to repair the damage caused by the crystallization of water in neural cells. Even if it took a hundred years to create the requisite nano-robots, eventually the resurrection and repair of the frozen dead would be possible. Today roughly a thousand people are signed up for cryonic suspension, or already are suspended. Cryonic suspension of one's head at Alcor, the leading cryonics firm, costs $100,000. Most cryonics patients pay for suspension with a supplemental insurance policy that is relatively affordable for middle-class Americans, although not for most people on the planet.

ORGANIZING THE PARTY OF LIFE

All this progress in forestalling death makes some people very nervous. The chair of the President's Council on Bioethics, the philosopher Leon Kass, has written widely against "unnatural" efforts to live longer, arguing that human life is given meaning by the certainty of death. Kass says, "The finitude of human life is a blessing for every individual, whether he knows it or not." In his new post he has opposed life extension and his Council report *Beyond Therapy* lays a large set of speculative concerns at the door of life extension. A Kass appointee to the PCB, Francis Fukuyama, writes in his book *Our Posthuman Future* that life extension will lead to rigid, risk-averse societies, ruled by slowly decaying seniors ogling the shrinking number of young bodies. Other critics warn that life extension will exacerbate overpopulation and the growing dependence of the retired on the shrinking working-age population.

All possible, but none of it different from the changes we have already seen in the last century. Because of longer lives and changing family structures in the twentieth century we had to invent nursing care systems and institute Social Security and Medicare, and some children have to wait until they are 60 to get their inheritance. People now live long enough to die slowly of Alzheimer's. But we've adapted, and we didn't tell anybody they weren't allowed any more medicine after the age of 80 because that was the natural limit to life. We didn't tell people in the developing world they weren't allowed to have medical technology because they were living too long given their overpopulation.

I'm sure the bioLuddite efforts to ban life extension treatments will be easily swept aside. But that doesn't mean we will all have access to these medicines. During the twentieth century, unequal access to medicine had a very marginal impact on the average life expectancy. In general the rich could not buy many extra years of life. In the coming decades, however, the affluent will be able to buy extra decades, and soon may buy extra centuries. Once we cross that threshold there will be an enormous pressure to make life extension medicine universally accessible. Every country will need to decide how much of their resources they will devote to life extension research and treatments, which treatments will be covered and which will be out-of-pocket, who will be able to take the treatments, and how the old-age pension system should be adjusted in response.

In reviewing the progress toward life extension, and the emerging political debate, science writer Ron Bailey predicts: "The defining political conflict of the 21st century will be the battle over life and death. On one side stand the partisans of mortality, who counsel humanity to quietly accept our morbid fate and go gentle into that good night. On the other is the party of life, who rage against the dying of the light and yearn to extend the enjoyment of healthy life to as many as possible for as long as possible."

I am confident that the party of life will succeed in the long run. But a lot of people may die unnecessarily as we fight that battle.

4

Getting Smarter

Research shows that people in the democratic industrialized countries have been getting smarter for at least the last century. Improving diets and health, the spread of education and more stimulating and complex lives all seem to make us smarter. But we appear to be reaching the limits of our ability to enhance intelligence without changing the brain itself, at least for the affluent. Fortunately technologies will soon be available that promise to make us more intelligent and improve our memories and concentration. If we are allowed to use them, and they are made widely available, they could improve the quality of our lives and make us dramatically better citizens.

EQUALITY, MODERNITY AND INTELLIGENCE

Francis Galton, Charles Darwin's cousin, was the father of population genetics. Galton also fathered the doctrine of "eugenics," the idea that the quality of the human race could be improved through selective breeding. Galton argued in his book *Hereditary Genius* that genius ran in some families of the British aristocracy, and must therefore be substantially genetic. The debate over the relative genetic and environmental contributions to intelligence has continued to this day, and both environmental and genetic factors are known to predict the variation of intelligence. However, one of the central predictions of the advocates of genetic determination of intelligence has not been borne out. Galton and the eugenicists predicted that there would be a gradual decline in intelligence since the lower classes and non-white races, who they believed to be less intelligent, had higher birth rates. In fact, intelligence has been rising.

In the 1980s, New Zealand political scientist James Flynn began to look at the questions used in intelligence tests in the twentieth century. The tests had changed, but the distribution of IQ hadn't since an IQ of 100 is by definition the average intelligence for a given population. But was today's 100 the same as last year's? Flynn found that IQ tests have been getting harder, and that intelligence has risen by roughly 30 points in the developed world, most of it since World War Two. Worldwide the average gain in intelligence between 1952 and 1982 was about 18 IQ points. On at least one measure, people who would have been in the top tenth of intelligence a hundred years ago would today be among the 5% *least* intelligent people in the population. Compared to previous generations, the number of people who score high enough to be classified as "genius" has increased more than twenty times. The "Flynn effect" has now been documented in dozens of countries, although most of the research has been in the democratic, industrialized world.

There is no widely accepted explanation for the Flynn effect, but the causes are assumed to be entirely environmental. Improved nutrition has contributed since there is a well-documented link between malnutrition and lower intelligence. The percentage of the population that can read has increased rapidly around the world, as has earlier, better and more widespread mandatory education. As family size shrank, each child received more attention. As people moved from culturally isolated farms to complex, stimulating cities, their brains were also stimulated and they developed more complex reasoning and problem-solving skills. Especially stimulating was greater exposure to all kinds of media, including books, magazines, radio, computers and television. While none of these factors appears to explain the Flynn effect by itself, each appears to contribute something.

These factors not only make us smarter, but keep us smarter. The stimulation of the brain provided by modern, educated, industrial life appears to help prevent the erosion of memory and reasoning ability among seniors. An active social life and intellectual activities like reading and playing bridge build a "cognitive reserve" that protects the brain in old age.

The environmental influences on intelligence contribute to a damaging feedback loop among poverty, poor nutrition, less education, ill health and lower IQ. People with sharper minds are more successful in life, make more money, are healthier and live longer. Conversely, hundreds of millions of children worldwide are malnourished and suffer lifelong intellectual deficits as a consequence. Some children go to schools with computer labs, some go to schools without paper and some don't go to school at all. Some people work at intellectually stimulating jobs while others work in boring, mindless jobs. Some seniors have had the education, and now have the resources and health care, that keep their minds sharp, and others aren't so lucky.

We have a long way to go in raising global living standards and access to education before we would reach the limits of how much intelligence can be nurtured and sustained through equality and environmental enrichment alone. But for affluent and middle-class people in the developed countries, we may be reaching the limits of how much intelligence we can achieve without genetic or pharmaceutical help. Eric Turkheimer, a psychologist at the University of Virginia, has shown that while environment is a strong predictor of intelligence for kids from poor families in the United States, genetics are the overwhelming predictor of intelligence for kids from middle-class and affluent backgrounds. On a scale of inheritability that ranged from 0, where kids' intelligence was not correlated with their parents' intelligence, to 1, where the kids' intelligence exactly matched their parents', poor kids scored 0.10 on inheritability and affluent kids scored 0.72. That means there is shrinking room for middle-class and affluent parents to help their kids be brighter through environmental enrichment alone.

HURRAH FOR RITALIN

Fortunately we have a growing number of technologies that extend our ability to concentrate, remember, discern patterns, solve problems and be creative. For instance, stimulant medications allow children and adults with attention deficit disorder (ADD) to focus and achieve more than they would otherwise. But as the number of kids diagnosed with ADD and successfully treated with stimulants has grown, so has the bioLuddite backlash. Francis Fukuyama condemns parents in *Our Posthuman Future* for drugging a generation of boys for just being naturally boyish. Fukuyama also suggests that we rob kids of self-esteem when they achieve with chemical assistance what they should have achieved through sheer willpower. This is like saying that someone who controls her blood sugar purely through diet should feel more virtuous than someone who relies on insulin.

There is no crisis of over-drugging of American kids with stimulants, much less kids anywhere else. The American Academy of Pediatrics estimates that as many as 10% of American kids have ADD and, in fact, only 3% of American kids are being prescribed stimulant medications for ADD. There are probably some kids being prescribed medication who don't really benefit from it, but many more who aren't getting it but need it. If anything, even more people could benefit from stimulants than could get a diagnosis of ADD. But giving a drug to someone without a clinical diagnosis just because it improves their performance crosses the bioLuddite's imaginary line between therapy and enhancement.

My family's experience with stimulants and ADD illustrates how difficult that line is to draw. When my son was in kindergarten, at our wonderful local university laboratory school with an almost 1-to-1 ratio of teachers and children, I came to dread the daily pickup routine. I would be debriefed about the daily fights, assaults and tantrums. Finally they told us our son was among the most difficult kids they had had to deal with, and that he would need a psychological evaluation to stay. The evaluation diagnosed ADD, as we had expected, and he was prescribed Concerta, a time-release version of Ritalin.

The first morning we gave him Concerta he went and sat on the couch before his ride to school and got very quiet. Concerned, we asked how he felt. He said, "Shhh . . . I'm thinking." He had never been able to slow down enough to notice his own thought process before, or to sit on our lap, or start to learn to read. But his behavior problems cleared up in kindergarten, and in elementary school he flourished. Now he is among the best readers and mathematicians in his class, while I'm quite sure he would still be struggling if it had not been for the medication. I'm pretty confident of that, since it took a combination of Ritalin and special education for me to learn to read in the third grade.

So is Ritalin an enhancement for my son who is just "naturally boisterous," or a therapy for a brain that needs to be adjusted to permit the child's fullest potential to shine through? Should we only give him enough Ritalin to bring him up to the class average, or is it OK to bring him to his fullest potential? If the latter, and if the drug is safe, why shouldn't all kids and adults who think they might benefit from it try it and see if it helps them achieve more, concentrate better and become more aware of their own thought process? Ron Bailey quotes Dartmouth neuroscientist Michael Gazzaniga: "What is the difference between Ritalin and the Kaplan SAT review? It's six of one and a half dozen of the other. If both can boost SAT scores by, say, 120 points, I think it's immaterial which way it's done."

Since he acknowledges there is no real line between therapy and enhancement, Leon Kass says that preventing a posthuman future requires preventing people from even getting close to being tempted to use medical technology to achieve "ageless and ever-vigorous bodies, happy (or at least not unhappy) souls, and excellent human achievement (with diminished effort or toil)."

MEMORY PILLS AND BRAIN FERTILIZER

Eventually we will slip off the Puritan chains and agree that each of us should be able to achieve our fullest, and decide for ourselves whether to use or forego technologies to get there. In the meantime, technologies of cognitive enhancement will all be developed, first as therapies for those with "impairments" and "diseases." Beyond the 5–10% who may be able to get prescriptions to treat ADD, another group set to benefit from cognition enhancement is senior citizens threatened by Alzheimer's disease and other dementias.

About half of all seniors who reach 85 have Alzheimer's disease. Since Alzheimer's is the number one cause of institutionalization in the United States, anything, be it education or a drug, that reduces the incidence of Alzheimer's would have enormous benefits for society and caregivers, not to mention for the individual facing that long, terrifying spiral. Research on the chemical precursors involved in memory consolidation, neural plasticity and synaptic transmission speed has led to clinical trials for more than forty drugs to treat dementia. The first of the new "smart drugs" on the market, Memantine, increases the availability of glutamate, a neurochemical necessary for communication between neurons. Memantine allows seniors with dementia to live independently for six to twelve months longer.

But what seniors diagnosed with dementia—and even people over 20 years old—really need is a way of reversing brain damage. Stem cell research has shown that the brain is able to repair itself by growing new neural stem cells well into the senior years. As research unravels how "neurotrophic" chemicals govern the growth of neural stem cells in the brain, we will begin to develop drugs that encourage brain self-repair.

GENE TWEAKS AND ELECTRODES

Even better than a neurotrophic pill would be a neurotrophic gene therapy that helped the brain to repair itself, or that enhanced intelligence in other ways. Researchers from the University of California at San Diego cultured skin cells from seniors with Alzheimer's disease, modified them with a gene to make them produce more nerve growth factor, and implanted them in the seniors' brains. Trials have shown that the technique reverses the atrophy of brain cells. Teams in Australia and New York are also treating Parkinson's disease with gene therapies that make missing brain chemicals.

Gene therapy brings us back to Galton and the eugenicists, who were half right about the inheritance of intelligence, although not about its relationship to race and class. Twins raised in different homes have almost exactly the same intelligence. In a study of hundreds of American child geniuses, Robert Plomin of London's Institute of Psychiatry found specific genes on chromosome 4 that only the genius children have. Plomin has also found that problem mutations on the same genes appear to give rise to all learning disabilities. These findings and other accumulating evidence give strong support to the idea that there are a finite number of genes that determine general intelligence, "g," and not just separate genes determining individual intellectual capacities such as memory, spatial visualization or verbal skills. People who inherit good genes for *g* tend to be better at everything. By manipulating genes, researchers have already been able to dramatically improve the intelligence of mice. So, although pessimists insist that intelligence is far too complex to be changed with gene therapy, it appears we will be able to tweak our own intelligence with gene therapies targeted at "g genes," and increase the intelligence of our children through germinal choice.

After cognition-enhancing drugs and gene therapies the most powerful human intelligence enhancer will be to directly connect the brain to computers. Researchers have been experimenting with two-way neural–computer communication by growing neurons on and around computer chips, or by placing electrodes in brains that are connected to computers, leading to computing prostheses for the brain. Completely paralyzed patients have been able to communicate by "typing" on computers connected to the motor neurons in their brain. The patient thinks about what it once was like to move the mouse, or their arm or leg, and the electrode picks up the neural signal and translates it through a computer into movement of a cursor. Combined with cochlear implants and electronic vision systems, or with nanoelectrodes directly to the cortex, patients could soon have direct input and output between their brain and computers.

Progress with brain chips and neural prosthetics is proceeding rapidly. The U.S. National Institutes of Health have a wide-ranging research program on neural prostheses, working on everything from control of prosthetic limbs to brain implants. The Defense Advanced Research Projects Administration is providing tens of millions of dollars to MIT, Duke and other universities through their "Brain Machine Interfaces Program" aimed at permitting soldiers to communicate with equipment, and each other, at the speed of thought. A lot of research is focused on creating new materials that can coexist with neurons indefinitely, and can be made into nanoscale electrodes.

Brain prostheses will also allow us to replace specialized brain structures lost to damage or disease. For instance, researchers at the University of Southern California have reverse-engineered the signals going into and out of the hippocampus, a part of the brain that affects memory, mood and awareness. They have designed computer chips to replace the hippocampus and are programming the prostheses with "neural network" software that mimics the way neurons make connections in the brain.

Several of the papers in the 2002 NBIC report (on the convergence of nanotechnology, biotechnology, information technology and cognitive science) extrapolate alternative methods for using nanotechnology to communicate with or augment mental functions in the coming decades. Recipients of "brainjacks" would be able to think directly into and directly call up calendars, address books, lyrics of songs, lines of dialogue for a play or foreign language dictionaries. With an Internet connection we could silently read our e-mail, surf the World Wide Web and carry on a mind-to-mind conversation with other brainjack users, all while pretending to pay attention in meetings. Some researchers suggest that these neuroprostheses could be perfected and available to the general public by 2030.

Beyond simply embedding computer chips in the brain, with very few sites for interaction, computer scientist Ray Kurzweil predicts that by 2030 we could have a dense latticework of nanobots that distribute themselves throughout our brains, communicating with every neuron, augmenting the brain's functions, feelings and thoughts. That would allow the brain to think faster thoughts; to multitask; to record and play back thoughts, dreams and feelings; and to switch seamlessly between immersive virtual reality and sensory reality. Most importantly a nanobot-pervaded brain would allow computing media to capture and recreate a dynamic model of each of our brains down to the last synapse. Since computers powerful enough to model human brains should be common in thirty years, those computer models may then be able to run software simulations of our brains and bodies. Presumably these backups of our minds, if switched on, would be self-aware and have an independent existence. This is the scenario known as "uploading." Uploaded people would be able to think at chip speeds, orders of magnitude faster than neurons, and be able to add additional hardware a lot easier than we will be able to add capacity to organic brains.

SMART CITIZENS, SMART DEMOCRACY

These scenarios of intelligence amplification with drugs, genes and machines have a lot of people spooked for a bunch of reasons that I address in later chapters. Many of the critics insist that these tools should only be used to fix cognitive diseases and disabilities, and not to enhance the normal brain. Critics from the Left worry that if intelligence amplification is available on the market it will exacerbate social inequality, and they are right. Securing our rights to become the most that we can be will require not only a fight for our individual rights to use technology to control our own brains, but also a fight to ensure universal access to intelligence-amplifying technology. Strong democracies should invest in their populations' intelligence not only through education, nutrition and health care, but also through ensuring that all citizens can take advantage of brain boosters.

In turn, as intelligence increases, democracy will be strengthened. The more intelligent the citizen, the more capable they will be at assessing their own interests, understanding the political process and effectively organizing. An increasingly intelligent citizenry will insist on a less authoritarian society, as John Stuart Mill suggested 200 years ago:

> From this increase of intelligence, several effects may be confidently anticipated. . . . They will become even less willing than at present to be led and governed, and directed into the way they should go, by the mere authority and prestige of superiors. . . . The theory of dependence and protection will be more and more intolerable to them, and they will require that their conduct and condition shall be essentially self-governed.

Democracy and technology have and will increase human intelligence, and increasing human intelligence will encourage liberty, democracy and the innovation of new technology.

Being Happier

Happiness is an illusive goal. But in democratic societies we are freer from mental and physical abuse at the hands of husbands, soldiers and unaccountable authorities than ever before. Medical technology is also freeing us from the discomforts of mental and physical illnesses. New pharmaceuticals and nanotechnologies will permit us happiness and freedom from pain that are currently unimaginable. Can we remain a democratic society if we all keep ourselves happy with drugs? Can we remain a democratic society if we refuse to allow people to control their own brains?

BEYOND PAIN

Long life, good health and smarts do not, unfortunately, make us happier just by themselves. Researchers have found, for instance, no correlation between happiness and intelligence, negative or positive. Perhaps this is because, as artificial intelligence theorist Eliezer Yudkowsky suggests, the capacity for both boredom and entertainment expands with our intelligence. Hobbies become boring as we mature, so we find new, more complex amusements and companions. The advantages of intelligence, such as greater success in life, more income and a lower likelihood of inflicting unnecessary misery on ourselves, are probably balanced by a greater awareness of the things we should be worried about and dissatisfied with. In a rough version of this balance researchers have shown that some of the genes that increase intelligence in mice also increase their sensitivity to pain. But neuromedicine may be the missing link between intelligence and happiness. Today human intelligence, in the form of technology, is about to make possible the elimination of pain and lives filled with unimaginable pleasure and contentment.

Three centuries ago the Scottish philosopher David Hume asked why we need pain. "It seems plainly possible to carry on the business of life without pain. Why then is any animal susceptible of such a sensation? If animals can be free from it for an hour, they might enjoy a perpetual exemption from it." This idea is offensive to many religions, which believe a life without pain would be somehow sinful and unnatural. BioLuddite critics, channeling this religious belief that life is supposed to be nasty, brutish and short, argue that life extension and happiness technologies will rob us of some of the ineffable quality of life and leave us bored and dehumanized.

The techno-optimists argue back that longer, healthier lives, with greater intelligence and less physical and mental pain, will also be richer and more fulfilling. I side with the optimists. Our most fundamental drive in life is to be happy, to reduce our pains and increase our joy and fulfillment. It seems obvious that the ethical goal for society should be to make life as fantastic for as many people as possible, not to valorize pain and suffering.

Before I get into the different ways transhuman technology and democracy will make us happier, we need to unpack some different kinds of suffering. There is physical pain, and the suffering caused by psychiatric illness. There is the ordinary suffering and happiness caused by our desires not being met by our lives. There is the angst of meaningless existence, the peace of spiritual contentment, the joy of "peak experiences." Emerging technologies are poised to make people happier in all of these ways.

So let's start with physical pain, which is a perfect illustration of the combined importance of technology and democracy in improving human lives. Pain is part nerve impulses and part the psychological suffering caused by powerlessness. A technology that addresses both aspects of pain has now been in use in hospitals for more than a decade: Patient-Controlled Analgesia (PCA). PCA machines allow patients to determine when and whether to inject themselves with a predetermined dosage of narcotic. The dosages are carefully rationed by the machine to prevent the patient getting high or taking an overdose. The irony of PCA is that patients using the devices feel *less* pain while using *fewer* narcotics than patients getting injections from nurses. Being able to self-administer reduces anxiety and alleviates much of the psychological suffering that accompanies the pain.

In terms of cybernetic pain control, recent research at Oxford University has found that electrodes implanted in the brain can relieve the chronic, disabling pain sometimes suffered after a stroke or head trauma. Neurosurgeon Tipil Aziz drills two holes in the skull and places electrodes deep inside the brain. As with PCA, the sufferer controls the electrical current. Although the surgery and device currently cost more than $40,000, since the therapy gets people off of continuous pain medication and back to work it is well worth it. Future nano–neuro interfaces will allow even more fine-tuned control of pain and pleasure.

Unfortunately our Puritan War on Drugs has made physicians leery of providing adequate amounts of narcotics to treat chronic pain and terminal patients' pain lest they be investigated by the Drug Enforcement Administration. After all, we wouldn't want anybody to get addicted to morphine, even if it is necessary for them to function or they only have six weeks to live. Study after study shows that most American patients dying with painful diseases, such as cancers, die in far more pain than necessary. The hospice movement and efforts to train doctors to provide "palliative" end-of-life care have so far had very limited success. Pain control is better in Europe because they dispense narcotics more freely. In a very real sense the repressive War on Drugs causes people pain without any demonstrable benefit for public health, and at the enormous cost of unnecessarily criminalizing millions of people. Giving people more control over their brains in the future certainly comes with risks, but the cost of criminalizing our control over our own brains is even greater.

People are also denied adequate amounts of narcotics because their doctors are afraid of being charged with physician-assisted suicide. Many people in the hospice movement have opposed moves to legalize physician-assisted suicide on the grounds that it will let doctors off the hook from having to better manage the treatment of the dying. But the evidence from Oregon and the Netherlands, the only U.S. state and the only country to have legalized assisted suicide, shows that giving people the right to control their own death actually improves the quality of their care. It encourages doctors to treat pain adequately, and to make end-of-life decisions openly and collaboratively instead of paternalistically. It turns out that just as many patients die from physician-administered drug overdoses in other European countries as in Holland, but without the discussion with the patient about their wishes that Dutch law makes possible. Just as with PCA, the quality of life of a patient with an emergency stash of pills improves because they have a sense of control over the end of life and the threat of unbearable pain. Having the option often relieves patients' anxiety enough so that most never make use of the pills.

BEYOND MADNESS

The second kind of suffering is caused by abnormal brain chemistries, the psychoses and depressions. Approximately 10% of the U.S. population suffers from a treatable mental illness at any one time, and as much as 20% will experience mental illness at some point in their life. In particular, about one in eight people will suffer from clinical depression, and 18 million Americans are suffering from depression at any given time.

Prozac, Paxil and the other selective serotonin reuptake inhibitors (SSRIs) have revolutionized the treatment of depression, and brought relief to tens of millions of people in the industrialized world. Millions more are being recognized as having mild depression, "dysthymia," which can also be treated with the new anti-depressants. Research shows that the SSRIs not only relieve depression but stimulate the growth of new neurons in the brain, helping to eliminate depression by changing the structure of the brain and protecting the brain from the damage caused by the stress of mental illness.

We are just at the beginning, however, of a revolution in the treatment of mental illness. A growing number of drugs, like Paxil and beta blockers, are helping people who suffer from disabling anxiety and shyness. People suffering from post-traumatic stress can use the drug propanolol to selectively dampen the emotional punch of their bad memories or forget them altogether. Anti-drug "vaccines" are cutting cravings to get drunk or high. Drug therapy is increasingly successful at treating obsessive-compulsive disorder.

Cybernetic implants and prostheses are also being used to treat mental illnesses. A crude contemporary example of future nano–neuro treatments for mental illness is the NeuroCybernetic Prosthesis System (NCP), originally developed to short-circuit epileptic seizures. The NCP sends electrical impulses to the vagus nerve in the neck from a generator tucked in a skin pocket under the collarbone. It turns out that these shocks stop not only seizures, but also depression, just like "shock treatments" or "electroconvulsive therapy" (ECT). Similar results have been found with Transcranial Magnetic Stimulation (TMS), which zaps a depressive's brain with powerful, focused magnets. TMS is as effective at treating depression as NCP and ECT, but without NCP's surgery or ECT's memory loss, seizures or anesthesia.

Depression and schizophrenia are both highly heritable, and teams are working to identify the genes that create them. The incidence of mental illness will therefore also be reduced by genetic screening, germline therapy on parents and embryos, and somatic genetic therapy on the sufferers themselves. Once the genes for mental illness are identified, drugs can be targeted at their mechanisms. Between gene therapies, better drugs and nano–neuro brain prostheses, mental illness will likely join cancer and aging in being completely preventable and controllable by the latter twenty-first century.

Also like physical illness, the treatment of mental illness is better in countries that respect individual rights and are egalitarian and compassionate toward the less fortunate. Today people in Western democracies can't be involuntarily committed unless they are a threat to themselves or others, and even then their rights as wards of society are far better respected than in the bad old days of the insane asylum. The poor are more likely to suffer severe consequences when they do become mentally ill, and the more egalitarian social democracies provide more generous social supports for the mentally ill. Far fewer mentally ill people in the developing world are diagnosed and treated than in developed nations. Addressing the unhappiness caused by brain chemistry requires not just universal access to the new pharmaceuticals but, equally important, investments in social reform and public welfare.

EVERYDAY JOY

As to the daily mismatch between our desires and our lives, new technologies promise to raise this everyday level of happiness. Like the tendency to depression, the ordinary level of daily happiness is partly genetically determined, as are personality traits like self-esteem that contribute to happiness. Research finds that happiness tends to be stable across one's lifetime, suggesting a genetically fixed happiness "set point." David Lykken and Auke Tellegen of the University of Minnesota found that identical twins have the same level of happiness 44% of the time, while nonidentical fraternal twins achieve the same level of happiness only 8% of the time. Dean Hamer of the National Cancer Institute, also studying twins, concluded that the heritability of "subjective well-being" was about 40–50%. The heritability of happiness, and the existence of the rare genetic mutation of hyperthymia, which gives its carriers an unfailingly sunny, positive disposition, suggests that there could be future drugs and gene therapies that jack our happiness set-point to its maximum without negative side effects.

Philosopher David Pearce thinks this is exactly what we should do. In his transhumanist utilitarian manifesto "The Hedonistic Imperative," Pearce argues that the explicit goal of democratic public policy should be to chemically increase citizen happiness to maximum limits. He proposes future drugs that jack the dopaminergic system to maximum production all the time, and marshals evidence that such drugs would not suppress motivation or have other negative consequences.

Francis Fukuyama strongly objects to Pearce's goal and brings up the drug soma from Aldous Huxley's *Brave New World,* which kept the populace from objecting to authoritarianism and "dehumanization." Fukuyama wants to draw a clear line between the use of drugs to treat depression and the use of drugs to boost the happiness of the ordinary person. He asks, "Could all that struggle in human history have been avoided if only people had had more serotonin in their brains?"

Such a dystopian outcome is possible if happiness drugs also make us stupid, lethargic and listless in addition to very happy. But, as Pearce suggests, a drug that made people more cheerful and optimistic would be as likely to give people the necessary hope and energy to improve their lives, to work on grand projects and change their world. There appears to be no contradiction between having an optimistic outlook and positive self-image, on the one hand, and a passionate commitment to civic engagement and social justice on the other. Perhaps these technologies will even offer the same "peak experiences" as the religious traditions, ranging in results from quietist spiritual bliss to full, joyous engagement with our lives and other people. The neurosciences are rapidly identifying the chemical pathways and brain structures that generate religious experiences. Andrew Newberg has been imaging the brains of people adept at prayer and meditation, and showing which brain structures are involved. Neurologist and Zen practitioner James Austin has outlined the neurophysiology of Buddhist meditation in *Zen and the Brain.* "Entheogen" researchers are working on the relationship of drugs to religious states. As we gain control over our brains, if we are allowed control over our brains, we will not only be able to control pain, mental illness and our daily level of happiness, but we will also have access to the "flow" experienced by athletes and yogis.

Freeing people from pain and depression and making them happy, cheerful and optimistic, may also encourage engagement with life, community and democracy. Democracy, in turn, is one of the best guarantors of general happiness of a population. Dutch sociologist Ruut Veenhoven has found, for instance, that the principal predictors of the average happiness of a country are how equal it is (democracy) and how prosperous it is (technology). Veenhoven created a world database of countries' happiness quotients, which he measured by the average level of subjective well-being on surveys multiplied by the country's life expectancy. The happiest nations by this measure were the northern European democracies, with Sweden, the Netherlands and Iceland at the top. The two best predictors of a country's Happy Life Expectancy were per capita income and a measure of equality between the sexes and the classes. Equality directly increases average happiness since each dollar in the hand of a poor person brings them more happiness than that same dollar in the hand of an affluent person. More equal distributions of wealth and generous social welfare benefits both lessen the suffering of the needy and the insecurity of the middle class. Veenhoven also finds that people are happier in societies that provide them civil and economic freedom, which corresponds to research showing that people who feel they are in more control of their lives are happier than those who think external events control their lives.

So just medicating the poor and powerless is unlikely to assuage their gnawing sense of injustice or tangible lack of stuff compared to the affluent. But there will probably be increasingly powerful ways to push the bliss button in our brains, and we will each have to learn to avoid the temptations to live with the lotus eaters. Although we should discourage addictions in one another, the War on Drugs is such an obvious failure and the right to control our brains so important that the decision about how happy and how motivated we should be should remain an individual decision.

John Stuart Mill said that "few human creatures would consent to be changed into any of the lower animals for a promise of the fullest allowance of a beast's pleasures. . . . It is better to be a human being dissatisfied than a pig satisfied." Mill also believed that when we are each given the opportunity to experiment and make decisions for ourselves, we will, in general, find a better mix of pleasures and pursuits than can be dictated for us by anyone else. In our transhuman future that freedom will include learning how to control our brains to achieve not just constant eyes-rolling-back-in-the-head bliss, but something more varied, interesting and meaningful, something that makes more use of our expanding intelligence, a level of happy, active intelligence that I firmly believe will lead citizens to a high level of civic and political engagement.

BARRIERS TO A BETTER FUTURE

The emerging technologies promise to fulfill our most basic desires, for strong, beautiful, long-lived bodies and brains capable of enjoying life completely and complexly. As I've repeated many times now in different ways, we can only fully enjoy the benefits of these technologies if we live in free, equal societies with strong, accountable democratic governments.

Fortunately the trends in the spread of democracy are almost as encouraging as Moore's Law of increasing computer power. One hundred years ago, 14% (6/43) of the world's countries were "democratic." In 1980, about 31% (37/121) of the world's countries were more or less democratic. By 2002, more than 63% (121/192) of countries were more or less democratic, including almost half the world's population. Democracy is more of a direction than a place, however, and no country in the world is as democratic as it should be. But whatever we call the various governments making the transition to multiparty elections and liberal civil societies, and looking past their temporary setbacks, the world has become progressively more democratic and more free over the last 300 years or so.

Yet unfortunately the world is not becoming more equal, and the persistent gap between the world's wealthy and poor is a serious threat to both democracy and access to technology. Most of the people reading this book are probably in the top 10% of the world's income earners, the 10% that 90% of all medical research is oriented toward. I'm confident that the transhuman technologies discussed here will eventually reach the rest of the 90% of the world. But the longer it takes to bring democracy and twenty-first-century technologies to the world's poor and powerless, the more lives that will be lost and stunted.

In *Our Posthuman Future* Fukuyama says the lesson to be learned from Huxley's *Brave New World* is that "we should continue to feel pain, be depressed or lonely, or suffer from debilitating disease, all because that is what human beings have done for most of their existence as a species." Then he ruefully notes, "Certainly, no one ever got elected to Congress on such a platform." Unfortunately I think some people will be elected with that platform. On the pretext of the uncertain consequences of technological progress, the bioLuddites will attempt to restrict access to these technologies by imposing authoritarian laws that forbid us to control our bodies and brains. In response, we need a global democratic movement that supports the creation of these new technologies, and that ensures that they are safe and made universally accessible in liberal democratic societies.

In the next couple of chapters I explain some of the origins of the bioLuddites' resistance to technological human enhancement, and outline an alternative vision of "cyborg citizenship" as the core of a movement for a transhuman democracy.

THE NEW
BIOPOLITICAL LANDSCAPE

From Future Shock to Biopolitics

The political terrain of the twentieth century was shaped by economic issues of taxation and social welfare, as well as cultural issues of race, gender and civil liberties. The political terrain of the twenty-first century will add a new dimension—biopolitics. At one end of the biopolitical spectrum are the bioLuddites, defending humanity from enhancement technologies, and at the other the transhumanists, advocating for our right to become more than human. Since biopolitics cuts across the existing political lines, it creates very odd coalitions. Either side can win public support depending on these coalitions and the lines they draw.

HALDANE'S REVOLUTION

John Burdon Sanderson Haldane's father was a professor of physiology at Oxford University in the late 1800s, and it was a foregone conclusion that JBS would follow in his father's footsteps. Indeed, at the age of 22 JBS was already becoming a respected biological researcher studying at Oxford. Then World War One intervened. Haldane joined the British army and by all accounts he was extremely brave and was wounded twice. But he said he was shocked that he enjoyed killing so much. The war left him profoundly disillusioned and he became convinced that human beings could and should use science and reason to improve every aspect of life. He would say later "the scientific point of view must come out of the laboratory and be applied to the events of daily life. It is foolish to think that the outlook which has already revolutionized industry, agriculture, war and medicine will prove useless when applied to the family, the nation, or the human race."

When Haldane returned to England and took a biology lectureship at Cambridge University he had become both a socialist and a dedicated evangelist for the possibilities of bettering humanity through genetics. Genetics was in its infancy and Haldane was one of its first and most important pioneers. Although Haldane flirted with eugenics as a young man, he became a vocal left-wing opponent of eugenics and its embrace of racism, classism and pseudo-science.

Nonetheless, Haldane continued to believe in the benefits of genetic science, and his 1923 speech "Daedalus, or Science and the Future" at Cambridge, which became a bestseller the next year, revealed the utopian hopes for genetics that motivated the contributions he made the rest of his life. In the essay he argued that all applications of biology provoke accusations that they are "unnatural."

But Haldane was convinced that great advantages would come from controlling our biology and from science in general. He projected a richer society, with abundant clean energy. He imagined a future democracy in which candidates for office would compete over who could make the next generation taller, healthier and smarter. Most controversially he imagined a future where "ectogenesis," gestating fetuses in artificial wombs, would be commonplace.

One of Haldane's close collaborators in British genetics was Julian Huxley. Julian's brother Aldous was greatly disturbed by the Daedalus essay and its enthusiastic reception among British intellectuals, and he wrote a scathing rebuttal in his classic *Brave New World*. In Huxley's horrific future artificial wombs are used by a totalitarian government to institute a rigid class system and expunge all lingering desires for monogamy or individuality. To this day those who propose to use technology to improve the human condition are accused of wanting to usher in a "brave new world."

This book is about the conflict between Haldane's optimism, that we could overcome our squeamishness about technology and use it wisely to build a better world, and Huxley's pessimism that biotechnologies will dehumanize and enslave us. Since the publication of Haldane's "Daedalus" and Huxley's *Brave New World* the debate about the political and ethical ramifications of transhuman technologies has been confined to the fringes of theology, bioethics, computer science and science fiction. Now, Leon Kass, Francis Fukuyama, Bill McKibben and a range of anti-technology groups have brought biopolitics to center stage. These bioLuddites argue that transhuman enhancement technologies are such a threat to "human dignity" and democracy that they must be banned. On the other side, the transhumanist descendents of Haldane are mobilizing to defend our rights to use reason and science to improve the human condition, to control our own bodies and to create a transhuman democracy safe for an increasing variety of citizens.

As we overcome our human limitations some people will be considered, or consider themselves, "posthuman." We will need to decide when and how humans should be allowed to become posthuman. Which forms of life should be forbidden, if any, and what does an entity need to pass from the status of property to person to full citizen— human DNA? intellect? communication skills? Does the right to control one's own body trump other people's distaste for the choices you make? I call this emerging complex of issues *biopolitics,* and this chapter explores how biopolitics is emerging as a new ideological dimension, ranging from reactionary bioLuddites to radical transhumanists, to complicate the landscape of twenty-first-century politics.

DEMOCRACY AND FUTURE SHOCK

We may expect . . . that a series of shocks of the type of Darwinism will be given to established opinions on all sorts of subjects. One cannot suggest in detail what these shocks will be, but since the opinions on which they will impinge are deep-seated and irrational, they will come upon us and our descendants with the same air of presumption and indecency with which the view that we are descended from monkeys came to our grandfathers. . . . [Most] profound will be the effect of the practical applications of biology.

—J. B. S. HALDANE,
"Daedalus, or Science and the Future"

I first read Alvin Toffler's *Future Shock* in high school. My father was an aspiring futurist and passed Toffler's 1970 bestseller along reverentially. Toffler proposed that the rate of technological change was increasing exponentially with each passing decade, and that growing numbers of people would suffer from "the shattering stress and disorientation that we induce in individuals by subjecting them to too much change in too short a time."

Toffler certainly had a point, a point familiar in sociology: People react badly when society changes too fast and they don't know what to believe in anymore. The eminent sociologist Emile Durkheim proposed in 1897 that rapid social change causes a widespread loss of certainty about the meaning and purpose of life, a condition he called *anomie,* which leads to a rise in suicide. Anomie was nineteenth-century French for "future shock." Social scientists returned to the idea of anomie to explain why people, disoriented by twentieth-century modernity, seek comfort in fascism or fundamentalism. Yet, although there is a rise in fundamentalism, most people have adapted to the accelerating rate of change, and adapted quickly.

My grandparents were all born within a couple years of the beginning of the twentieth century. During their lives they saw the United States turn from a primarily agrarian society into a fully industrial and then post-industrial society. They lived through the introduction of women's suffrage, the banning and legalization of alcohol, the rise of communism, the rise and fall of fascism and the spreading use of automobiles, television, contraception, computers, space travel, nuclear weapons, invitro fertilization and genetic engineering. In 1900, the average life expectancy in the United States was 47 years; by the end of the century it was 77 years. At the turn of the century, 80% of households were headed by married couples with children, and by the end of the century fewer than half were. Yet my grandparents, like most people of their generation, adapted to modern life with apparent ease, with resilience and pragmatism, without feeling overwhelmed or needing to escape into a future-rejecting belief system.

Even though the pace of change keeps accelerating, Toffler's concerns about "future shock" now seem overblown. Of course, some of the changes that Toffler predicted didn't come about, at least not as quickly as he proposed. But many of his predictions did come to pass. Toffler predicted the adoption of gay marriage, for instance, which is rapidly spreading in Europe and Canada, and will hopefully spread from Massachusetts across the United States. Perhaps the current violent reaction to gay rights and gay marriage is an example of Toffler's future shock. If so, what alternative do we really have but to press for equal rights anyway? Is social discomfort really a warrant for denying people their freedom? Having Rosa Parks sit in the white section of the bus in Montgomery, Alabama, certainly was disorienting and disturbing for many white Southerners. The education of women was deeply troubling to many tradition-minded men in nineteenth-century Europe, as it still is in parts of the twenty-first-century Middle East.

But eventually people adapt. People who felt that the institution of marriage and family would be fundamentally wrecked and that social chaos would ensue once gays lived openly next door have discovered that life goes on. Most Europeans and most young Americans now accept that sexual orientation is irrelevant to whether a person should have equal rights. Granting gays and lesbians legal equality has been a hard-fought social struggle, and it is not over. Gays and lesbians are still being harassed and killed, and the struggle to win full equality in all fifty U.S. states will take another generation. But eventually the logic of democracy will make laws against gay marriage seem as curious and wrong-headed as laws against interracial marriage are seen today.

As disorienting as the principled respect for individual liberty can sometimes be, democratic societies also empower majorities to put on the brakes when things begin to move too fast or in the wrong direction. Toffler argued in *Future Shock* that futurology needed to be woven into democracy so that the public could anticipate and control the consequences of new technologies. Anticipatory democracy as envisioned by Toffler, and practiced around the world, both slows the pace of change and adapts the public to it by giving voice to their concerns and a sense of empowerment over social change. In democracies people in the minority can express their complaints and organize in like-minded associations and communities. People living under authoritarianism are more likely to feel they are being steamrolled by the future—especially today by the capitalist American future—and react with authoritarianism and violence of their own.

In the last thirty years many governments have responded to Toffler's call for "anticipatory democracy" by appointing futurist advisory commissions. They have had mixed success. For instance, President Jimmy Carter appointed an interdepartmental commission to craft the Global 2000 Report. Like the pessimistic Club of Rome studies that projected global apocalypse by 2000 based on Malthusian modeling, the massive Global 2000 report projected imminent environmental degradation, shrinking resources, vast increases in poverty and widespread starvation, unless drastic efforts were made at population control and sustainable technology. The report was delivered to President Ronald Reagan in 1981 and promptly ignored.

In the twenty years since that report, a combination of technological advances, incremental environmental policies and successful population control efforts have kept the worst of its projections from coming true. This effort at anticipatory democracy failed because our efforts at anticipatory social policy were incrementally effective. The environment is still being assaulted, stores of nonrenewable resources are still shrinking and global inequality continues to widen. But the present global crisis is not of the apocalyptic dimensions envisioned in the late 1970s. The Global 2000 Report is an important cautionary example; bodies charged with anticipating the future can wildly overestimate the risks of techno-logical progress, and underestimate our social and technological capacity to address these risks.

President Carter also commissioned the first executive-level American bioethics commission, The President's Commission for the Study of Ethical Problems in Medicine and Biomedical and Behavioral Research. By 1982, the Commission had issued ten reports that have had far-reaching impacts, on topics such as the definition of death, informed consent, genetic screening, health care access, life-sustaining treatment, privacy and confidentiality and genetic engineering. The brain-death laws that the commission drafted, which I discuss later, have been adopted by most states and most of the industrial world. In the case of Carter's bioethics commission, and similar bodies in many countries, anticipatory democracy helped adapt laws and values to technological change, tempering our future shock.

BIOETHICS AS BIOPOLITICS

After the Carter commission's mandate expired, Congress and Presidents Reagan and Bush were unable to appoint bioethics advisory bodies because of opposition from the anti-abortion lobby. Bioethics, it turns out, is just an especially philosophical branch of politics. Bioethics is proto-biopolitics being fought out in rarefied circles—academia and think tanks—before it breaks onto the popular consciousness and becomes biopolitics proper. Even the core principles of contemporary bioethics—autonomy, justice and beneficence—are direct corollaries of the French revolutionary slogans of liberty, equality and solidarity. Unfortunately those aren't the only values shaping bioethics.

In the 1970s, the focus of most bioethicists' attention was on protecting patients from unethical scientific research and overly aggressive applications of end-of-life care, protecting the public *from* science and technology rather than securing their rights to it. Bioethicists also began to raise questions about the dangers of cloning, invitro fertilization and genetic engineering. There were occasional courageous provocateurs, such as Joseph Fletcher, who argued that humans have a right and obligation to control their own genetics. But as bioethics matured it became clear that professional bioethicists gained far more traction by exacerbating the public's Luddite anxieties than by assuaging them. How often would a reporter call a bioethicist for comment if all he got was "That sounds like a wonderful medical advance, and it should help a lot of people—no problem"? If cloning is really just the creation of delayed twins, and not a profound threat to everything we hold dear, who is going to fund bioethics conferences to address it, and empower bioethicists to forbid scientific research into cloning? Rather than using their core democratic principles to ease our future shock and secure our rights to technology, many bioethicists became propagandists of future shock.

Today many bioethicists, informed by and contributing to the growing Luddite orientation in the social sciences and humanities, start from the assumption that new biotechnologies are being developed in unethical ways by a rapacious, patriarchal medical–industrial complex, and will have myriad unpleasant consequences for society, especially for women, the poor and the powerless. Rather than emphasizing the liberty and autonomy of individuals who may want to adopt new technologies, or arguing for more equitable access to new biotechnologies, bioethicists often see it as their responsibility to slow the adoption of biotechnology. The appointment of arch bioLuddite Leon Kass as the chair of President Bush's Bioethics Commission (PCB) in 2001 marked the crest of the bioLuddite influence in bioethics. It has, in turn, motivated non-Luddite bioethicists to become more vocal in opposition.

While President Bill Clinton's Presidential Bioethics Advisory Commission was broadly representative of academic bioethics, the political design of Bush's PCB was quite naked. The elevation of Kass, a man who has opposed every intervention into human reproduction from invitro fertilization to reproductive cloning, was a shrewd political payoff to the right-to-life movement, since he could clothe opposition to research on embryos in secular, albeit pre-modern, philosophical language. Kass in turn stacked the committee with Christian conservative bioethicists, such as Mary Ann Glendon and Gilbert Meilander, and conservatives with little or no connection to academic bioethics, such as Robert George, Francis Fukuyama, James Q. Wilson and Charles Krauthammer. The executive director for the PCB was Dean Clancy, a former aide to far-Right Texan Republican leader Dick Armey. After leading the PCB to recommend the criminalization of the use of embryos and embryo cloning in research, Kass focused the PCB on opposition to human enhancement, from psychopharmaceuticals to life extension, resulting in the 2003 opus *Beyond Therapy*. This report reflects Kass's well-known philosophical "yuck factor" or "wisdom of repugnance" approach: If a practice is scary or repugnant, that is sufficient grounds to ban it. Kass and the bioLuddites want to create new laws and institutions with the mandate to forbid technologies on the basis of these vague anxieties.

One might mark the opening salvo of the liberal, transhumanist-inclined bioethics resistance to Kassism as the essay "Leon the Professional" by philosopher Glenn McGee, which prefaced a special 2003 volume of the *American Journal of Bioethics* devoted to the ethics of human–animal transgenic experimentation. McGee hits the central tenet of Kassist bioLuddism: "If we get past the 'yuck'—as is suggested by more than half the contributors to the collection—we find that engineering of humans is not only ubiquitous and a function of ordinary human life as well as high-technology science, but also that the rules for avoiding 'yuck' are a mere matter of faith themselves in the articles of a flimsy new kind of neoconservative natural law theory. And perhaps we are better off yucky but complicated than in the clean, well-lit spaces of the illusory safety of a 'nature' that doesn't really exist."

JEREMY RIFKIN AND ODD BEDFELLOWS

I have no doubt that in reality the future will be vastly more surprising than anything I can imagine.

—J. B. S. HALDANE

Although Kass and the Christian Right are the most influential seg-
ment of the emerging bioLuddite bloc, they have increasingly been
joined by people from across the political spectrum, including from
the Left. The principal far-sighted strategist who has brought the Left
flank of bioLuddism into alignment with the Right is the veteran activist
and writer Jeremy Rifkin.

In the 1960s and 1970s, Rifkin was an anti-war organizer and socialist
activist. But in the late 1970s, Rifkin had a vision that the terrain of future politics would be fundamentally transformed by biotechnology in
the same way that steam power and electricity had created fundamentally new political and economic orders. In 1977, Rifkin went on to start
his think tank, the Foundation on Economic Trends, dedicated to throw-
ing roadblocks in the way of biotech. Rifkin named his nemesis *algeny,*
"the improvement of existing organisms and the design of wholly new
ones with the intent of perfecting their performance." But, for Rifkin,
algeny was also "a way of thinking about nature, and it is this new way
of thinking that sets the course for the next great epoch in history."

Rifkin quickly discovered the importance of alliances with the religious Right built on their shared critiques of algenic hubris. In one campaign Rifkin organized disgruntled former surrogate mothers and shep-
herded them around the United States to help pass laws banning
surrogacy contracts. Rifkin used that campaign to build ties between
Catholic conservatives who supported the Papal ban on surrogacy and
feminists uneasy with "uteruses for hire."

One of the issues that Rifkin sees as a clear and present danger is the
crossing of species barriers using recombinant genetic engineering, a
point that resonates with Christians concerned about humans "playing
God." So Rifkin reached out to religious groups arguing that these recombinant techniques were not only dangerous capitalist imperialism,
but also were violating God's plan for separately created species and robbing life of its "sacredness." In 1995, Rifkin announced that leaders representing more than eighty different religious groups had signed his
"Joint Appeal Against Human and Animal Patenting," which read: "We,

the undersigned religious leaders, oppose the patenting of human and animal life forms. We are disturbed by the U.S. Patent Office's recent decision to patent human body parts and several genetically engineered animals. We believe that humans and animals are creations of God, not humans, and as such should not be patented as human inventions."

In 2001, a heated battle raged between a broad coalition defending medical researchers' use of cloned embryos to generate stem cells and the right-to-life movement and new Republican president who favored a ban on use of embryonic stem cells. In the midst of this battle Rifkin sent out a petition to support a ban on "cloning" to prominent left-wingers and feminists. His petition had neo-conservatives William Kristol and Francis Fukuyama as co-signatories, and Rifkin said he wanted to unite the social conservative and liberal Left camps around a shared opposition to "cloning." Although many prominent progressives signed on, a good number thought the statement was calling only for the criminalization of reproductive cloning, not for the criminalization of medical research using embryos. Many said they would not have signed if they had understood that they were backing Republican legislation to criminalize medical research.

The petition lured in the inattentive progressives because it played on the vague bioLuddite assumptions that often substitute for progressive political analysis: Biotechnologies only serve the nefarious ends of their corporate manufacturers; embryos and the natural order need protection from "commodification" and "instrumental values." The letter read: "We are also concerned about the increasing bio-industrialization of life by the scientific community and life science companies and shocked and dismayed that clonal human embryos have been patented and declared to be human 'inventions.' We oppose efforts to reduce human life and its various parts and processes to the status of mere research tools, manufactured products, commodities and utilities."

Rifkin is quite clear about the importance of his odd coalitions to the coming "fusion biopolitics." In a 2001 article titled "Odd Coupling of Political Bedfellows Takes Shape in the New Biotech Era," Rifkin writes:

"The Biotech Era will bring with it a different constellation of political visions and social forces, just as the Industrial Age did. The current debate over embryo and stem cell research already is loosening the old political allegiances and categories. It is just the beginning of the new politics of biology." Rifkin is right about the new biopolitics, and his successes build on the commonalities of bioLuddism on the Left and Right. But the transhumanists are building some odd coalitions as well.

POLITICAL IDEOLOGIES

I can foresee the election placards of 300 years hence . . . "Vote for Smith and more musicians," "Vote for O'Leary and more girls," or perhaps finally "Vote for Macpherson and a prehensile tail for your great-grandchildren."

— J. B. S. HALDANE,
"Daedalus, or Science and the Future"

Most people don't describe their beliefs with an "-ist" at the end, and the number who do has been shrinking. More and more people are becoming independents, tugged back and forth by the small percentage of the population who are political activists and ideologues. But people do know they want health, longevity, security and prosperity for themselves and their kids. When push comes to shove very few are eager to embrace sickness, disease, mental decline and early mortality in order to avoid hubris or preserve their "humanness." So the constituency for a political ideology that promotes human enhancement includes the vast majority of the planet's people, rich and poor, religious and secular.

On the other hand, everyone also wants to avoid a future with apocalyptic threats, extreme inequality and wrenching social conflicts, so the fearmongering bioLuddites and religious fundamentalists may also be able to win majorities for bans on human enhancement. The fear-mongers certainly have the head start. But one or more brands of techno-optimism will inevitably emerge to organize people's hopes. What flavors will emerge? Why do some flavors win out over others?

Political ideologies are simplifying rules of thumb that allow people to make sense of thousands of choices, identities and interests. Ideologies crystallize out of the swirling mess of ideas in the heat of specific conflicts. We find ourselves on one side or the other of a picket line, and try to figure out why we are on one side and other folks are on the other. People on each side have different theories and visions of how to proceed, lines they want to draw for who belongs in which corner. The explanations and visions that win, partly by chance and partly by the hard work of their advocates, become ideologies, and they take on a life of their own. They need a little internal logic and consistency, but not a lot. Different ideologies can win and shape struggles in any given situation.

When about sixty members of the small Students for a Democratic Society (SDS) gathered in Michigan in 1962 there was little indication that their Port Huron Declaration would become one of the defining documents of the North American New Left, or that SDS would grow to 100,000 members and more than 400 chapters. The Declaration advocated a "participatory democracy" that would transcend the stalemate of capitalism versus communism. There were other thinkers, organizations and activists at the time who could have become leaders of the student, anti-war and civil rights movements, all with their own theories and manifestoes. In other countries variant ideologies dominated the 1968 student movement niche, such as the anarcho-surrealist Situationists in France, the ultra-nationalists of Turkey or the communist sects in Japan. But through a combination of luck and extraordinary organizational skills, the early leaders of the SDS helped make "participatory democracy" a central part of U.S. New Left ideology.

Future biopolitical ideologies are emerging in the same way. Rifkin or Fukuyama may lose out to a more successful species in the bioLuddite niche, and the transhumanist bloc may be organized by people who don't call themselves "transhumanist." But the ideological schism between bioLuddites and transhumanists is emerging, and aspiring organizers are jockeying to define the coming battles in terms that favor their place in the ideological spectrum.

MAPPING BIOPOLITICS

In the last century you could pretty much accurately place someone polit-ically by where they stood on two basic sets of political issues: economics and culture. Economic conservatives aren't interested in reducing inequal-ity, and don't care for the welfare state, trade unions, taxation, business regulation and economic redistribution. Economic progressives want peo-ple to be more equal, and generally favor all of these government mea-sures. Cultural conservatives are generally nationalistic, ethnocentric, reli-giously conservative and skeptical of women's equality, sexual freedom and civil liberties. Cultural progressives are generally secular and cos-mopolitan, and supporters of civil liberties and minority and sexual rights.

Where people and parties fall out on each of these axes predicts their positions on other issues on that axis, but not how they feel about issues on the other axis. The issues within each axis have some ideological and practical consistency that hold them together. People who are tolerant of changing gender roles and women's rights are also more open to changing sexual mores such as gay rights. Opponents of social welfare are more likely to support lower taxes. But knowing how someone feels about women wearing pants doesn't tell you how they feel about right-to-work laws.

The terrain that these two axes create, as shown in Figure 6.1, allows us to map out how and why alliances form and shift. The economic in-terests of white working-class people have generally led them to the upper half of the box, economic progressivism, while their educational backgrounds have made them more culturally conservative, leaning them toward the left-hand side of the box. So the "natural" politics of the working classes is the culturally conservative populism of Huey Long or Pat Buchanan in the United States, or a Juan Perón of Ar-gentina. Trade unions and social democratic parties have generally been led by well-educated cosmopolitans, however, people trying to build al-liances with the culturally liberal middle classes, so they pull in working-class support for the upper-right-hand "social democratic" corner. When

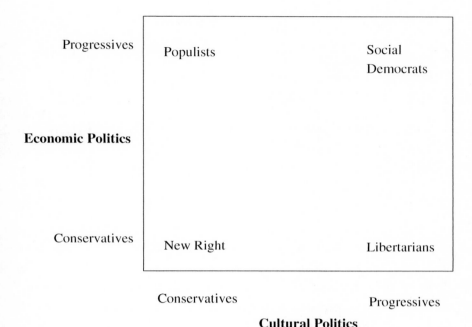

FIGURE 6.1   The Political Terrain of the Twentieth Century

working people stop believing their economic interests are represented by the social democrats, their distaste for immigration, gay rights, affirmative action and abortion allows them to be pulled back toward the religious Right in the lower-left-hand corner, anchored by the conservative churches that workers and the poor often attend.

Gender is also tied to political leanings, with men tending toward cultural and economic conservatism. Economically, men tend to favor the cowboy individualism of the free market, whereas women are more supportive of the nurturing welfare state. Culturally, men are less supportive of women's rights and sexual diversity. So men tend toward the New Right corner and women toward the social democrats.

What is really interesting and new about twenty-first-century biopolitics, as illustrated in Figure 6.2, is that this new, third dimension sticks straight out of the two-dimensional map. This gives rise to Rifkin's odd Left–Right coalitions.

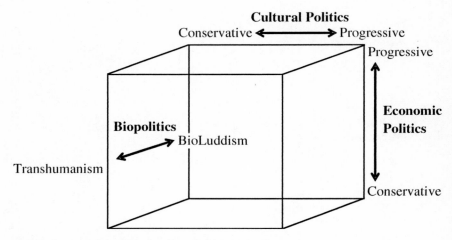

FIGURE 6.2   The Political Space of the Twenty-First Century

As biopolitics crystallizes out of issues and political struggles from treatment of the comatose and stem cells to genetically modified food and cloning, activists, parties and ordinary citizens are nudged toward biopolitical consistency. In the next chapter, I talk about five technologies, such as artificial wombs and the intellectual enhancement of animals, that will force people to take a side on one of the fundamental biopolitical disagreements between bioLuddites and transhumanists: whether citizenship should be tied to psychological personhood or genetic humanness.

As with the economic and cultural axes, there are already constituencies that lean toward one end or the other of the biopolitical spectrum. Looking just at techno-optimism, one element of the biopolitical schism, the constituencies that feel most positive about human reproductive and genetics technology tend to also feel optimistic about the space program, nanotechnology and other technologies. The reverse is not always the case though; more Americans are optimistic about the benefits of *non*-genetic technologies, such as space and computing, than genetics. In other words, attitudes toward genetic and reproductive technology are a more sensitive barometer of someone's techno-optimism. If you are optimistic about genetics you have put future shock behind you.

A large 2000 National Science Foundation (NSF) survey of American attitudes toward science and technology asked whether the benefits of genetic engineering would outweigh its dangers. Four out of ten Americans thought genetic engineering's benefits would outweigh the costs, while 28% thought the benefits and costs would be balanced, and 32% believed costs will outweigh benefits. In another national survey of Americans in 2000, conducted by the Public Policy Research Institute at Texas A&M University, 53% said that genetic engineering would "improve our way of life in the next 20 years," and 30% said it wouldn't. In these and other surveys the majority of American respondents have been in favor of access to invitro fertilization, genetic therapy and genetic screening and abortion for disabled fetuses. Still, only a minority are in favor of the "enhancement" technologies. About a quarter of Americans favor genetic enhancement and "designer babies" and about one in ten favor legal reproductive cloning.

On the other end, hard-core bioLuddites appear to comprise only about a quarter to a third of the population. About a third of Americans consistently oppose stem cell research and therapeutic cloning. In a 2002 survey of Americans conducted by the Genetics and Public Policy Center at Johns Hopkins University, a quarter to a half were opposed to prenatal selection and invitro fertilization. So, depending on the issue, between 10% to a majority might end up with the transhumanists, and 25% to a majority might end up with the bioLuddites.

The dynamics of the bioLuddite–transhumanist split vary around the world. Europeans, still spooked by Nazi eugenics, are more negative toward all reproductive technology and genetic engineering. Asians, on the other hand, are generally more positive than Americans toward these technologies. In a 1993 survey of more than 900 Indians and Thais, a majority in both countries supported genetic enhancement of physical characteristics and intelligence, and even making people more ethical. China has a law specifically directed toward improving genetic health, and Singapore's government has invested in efforts to encourage college graduates to have more children.

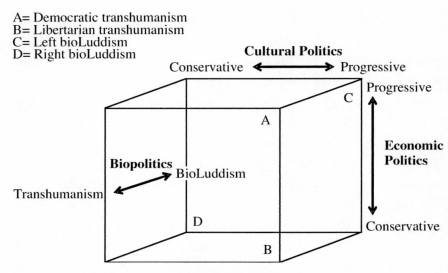

FIGURE 6.3   Ideological Positions in the Twenty-First Century

Generational change and rising educational levels and secularism worldwide appear to be on the side of techno-optimism since it is more common among the young, the college-educated and the secular. Men also lean toward techno-optimism. BioLuddism is more common among older people, the less educated, the more religious and women. In the 2000 NSF survey, men believed genetic engineering's benefits would outweigh costs 11% more often than women did (45% to 34%), and college graduates were more optimistic than those with high school diplomas by 11% (48% to 37%). A 2001 Gallup poll on animal cloning found that 56% of those with postgraduate education said animal cloning should be allowed, compared to only 19% of those with a high school diploma. Women were much more likely to oppose animal cloning than men (74% of women to 53% of men opposed).

Given these predilections, it is natural that the first transhumanist organizations have been more attractive to affluent, well-educated men, many with libertarian leanings down in corner "B" in Figure 6.3. But there are ways to craft the transhumanist message and coalition to appeal to and address the interests of women, working people and other constituencies suspicious of technology for one reason or the other. This book is an argument for building a new radical biopolitics around corner "A" where social democracy meets transhumanism. I call this corner "democratic transhumanism."

In Chapter 7 I outline the principal argument that divides the bioLuddites and the transhumanists: the conflict between personhood-based "cyborg citizenship" versus "human-racism." Then in Chapters 8 and 9, I describe other arguments and ideologies that make up the tapestry of bioLuddism, weaving together the progressive bioLuddites at "C" and the religious New Right bioLuddites down in "D." In Chapter 10 I discuss the history of the ideas that make up transhumanism, to show that it is a historical accident that most transhumanists today are found down in the libertarian corner "B." In the last part of the book I outline the democratic transhumanism that is emerging and sketch out the nature of some of its agenda.

7

Cyborg Citizenship

The most important disagreement between bioLuddites and transhumanists is over who we should grant citizenship, with all its rights and protections. BioLuddites advocate "human-racism," that citizenship and rights have something to do with simply having a human genome. Transhumanists, along with most bioethicists and the Western democratic tradition itself, believe citizenship should be based on "personhood," having feelings and consciousness. The struggle to replace human-racism with personhood can be found at the beginnings and ends of life, and at the imaginary lines between humans and animals, and humans and posthumans. Because they have not adopted the personhood view, the human-racists are disturbed by lives that straddle the imaginary human/non-human line. But technological advances at each of these margins will force our society in the coming decades to complete the trajectory of 400 years of liberal democracy and choose "cyborg citizenship."

BETWEENNESS AND THE "YUCK FACTOR"

The chemical or physical inventor is always a Prometheus. There is no great invention, from fire to flying, which has not been hailed as an insult to some god. But if every physical and chemical invention is a blasphemy, every biological invention is a perversion. There is hardly one which, on first being

brought to the notice of an observer from any nation which has not previously
heard of their existence, would not appear to him as indecent and unnatural.

—J. B. S. HALDANE,
"Daedalus, or Science and the Future"

In this age in which everything is held to be permissible so long as it is freely
done, and in which our bodies are regarded as mere instruments of our au-
tonomous rational will, repugnance may be the only voice left that speaks up
to defend the central core of our humanity.

—LEON KASS,
"The Wisdom of Repugnance"

Most bioethicists are skeptical of Leon Kass's idea that taboos and gut reactions are the path to wisdom. After all, as Haldane says, the progress of civilization has often involved overcoming our gut feelings that some things were obviously wrong or immoral. Until the French Revolution prohibitions against performing autopsies stymied progress in understanding anatomy. What, Haldane asks, must the tribe have thought of the first person to suggest drinking the milk from a cow?

We are correctly repulsed by actual violence and oppression against real people, by murder and child abuse. But often our gut feelings are simply indefensible prejudices about other peoples' choices. One of the greatest "yuck factors" for twentieth-century conservatives was "race-mixing." Whites were whites, blacks were blacks and mixed-race children were abominations. Then the logic of the equality of persons, regardless of skin color, broke through the laws against race-mixing and slowly the feelings of revulsion have disappeared. As bioethicist Art Caplan said, "The problem with repugnance and fear-and-trembling ethics is they are good starting points but bad ending points. . . . If intuition is the last word, then African-Americans are at the back of the bus, women and people who have no property aren't voting, and we still have slaves."

Today the "abominations" enraging the bioLuddites are lives that blur the ancient cognitive categories that separate animals and humans,

humans and machines, the living and the dead, the real and the artificial, men and women, the young and the old. In the twenty-first century, transhuman technologies will not only let us live longer, be smarter and have more control over our emotions and our bodies. They will also permit us to clone, to mix human and animal DNA and genetically modify our bodies for aesthetic reasons. We will incorporate computers into our bodies and brains, and simulate human brains in computers. Day by day our fellow citizens will become rapidly more strange, and we will need new categories and a new understanding of democracy to make new sense of the world.

In the 1930s, anthropologists began to describe cultural myths and belief systems as if they were architectural structures, with sets of symbols that were weighted and balanced against one another in a giant latticework, meaningless outside each culture but absolutely essential inside them. Anthropologist Mary Douglas used this method to understand *kashrut,* Hebrew dietary laws, in her 1970 book *Purity and Danger: An Analysis of the Concepts of Pollution and Taboo.* In Leviticus and Deuteronomy the Israelites were specifically instructed to eat ox, sheep, goats, gazelle and deer. But they were forbidden to eat the "abominable" animals: camels, rabbits and pigs. Douglas proposed that the logic of *kosher* (clean) and *treif* (unclean) was that clean animals either had cloven hooves and chewed a cud or did neither, as with birds. But camels chew cud and don't have cloven hooves. Rabbits appear to chew cud and are not cloven-hoofed. Pigs are cloven-hoofed but do not chew cud. So these confusing animals are unclean "abominations." The same logic can be seen at work when Leviticus condemns men who "lie with other men as they would with a woman," and in the moral codes of many other cultures. Things that lie between the understood categories of the cultural world are often considered abominable and frightening.

Yuval Levin addressed the betweenness taboo in his article "The Paradox of Conservative Bioethics" in the premier issue of the bioLuddite journal *New Atlantis.* Levin, a senior research analyst for Kass on the U.S. President's Council on Bioethics, said the agenda of conservative,

Kassite bioethics is to argue for bioLuddite bans, against the advancing tide of individual freedom, based on nothing more than taboos. Echoing Mary Douglas, Levin defines taboos as "the transgressing of a boundary, or a mixing together of things that ought to be kept separate. Taboos stand guard at the border crossings between the realm of the properly human and those of the beasts and the gods." We know how humans should be treated, and we know how non-humans should be treated. But what should we do with the creatures in between?

Levin goes on to warn conservatives that when they debate how to regulate abominable biotech, instead of just call for it to be banned, they normalize the idea of a transhuman future. The central problem for conservatives who defend taboos, says Levin, is that democracies do not forbid things just because they are taboo. Democratic societies require compelling reasons to interfere with people's liberty, and religious prejudices and gut feelings don't count. According to Levin, once conservatives begin to defend their taboos in democratic debate, they win battles but lose the war.

HUMAN RACISM

Levin is right. And democratic debate will defeat bioLuddite bans against human enhancement for a second reason. Not only are democratic societies expanding personal liberty, they are increasingly specific about what kinds of lives do and don't get to enjoy the benefits of that liberty. The bioLuddites advocate human-racism, that some ill-defined "humanness" is the green card to citizenship. Letting in non-humans or posthumans, or excluding embryonic or brain-dead humans, threatens the very fabric of the human-racist concept of the democratic polity. In *Our Posthuman Future*, Francis Fukuyama argues that any effort to change basic human biological traits threatens the entire system of human rights since these traits are the only common bond we share. Left-leaning bioethicists George Annas and Lori Andrews call for an international treaty to ban cloning and inheritable genetic modifications in their article "Protecting the Endangered Human": "Cloning and inheritable genetic alterations can be seen as crimes against humanity of a unique sort: they are techniques that can alter the essence of humanity itself (and thus threaten to change the foundation of human rights) by taking human evolution into our own hands and directing it toward the development of a new species, sometimes termed the 'posthuman.'. . . Membership in the human species is central to the meaning and enforcement of human rights."

In an earlier law review article Annas makes clear that cloning and germline modifications are only the first items on the list of species-altering technologies he would want banned. Annas would also support international treaties to ban "human/machine cyborgs, xenografts (transplants of animal organs), artificial organs, embryo research and brain alterations. These could all fit into a new category of 'crimes against humanity' in the strict sense, actions that threaten the integrity of the human species itself." With this expansive definition we can imagine makers of prosthetic limbs and cochlear implants being thrown into prison in the Hague for having created posthumans.

These well-intentioned advocates of human-racism, from Annas and Andrews on the Left to Kass, Fukuyama and the Christian Right, have missed the real insight of Western democracy: Citizenship is for *persons,* not humans. Persons don't have to be human, and not all humans are persons. To create a transhuman democracy we will have to establish a new definition of citizenship, a "cyborg citizenship," based on personhood rather than humanness. With cyborg citizenship we can deal with the scary boundary-crossers, the cyborgs, the animal–human hybrids, the genetically engineered kids, the clones and the robots. We can add some more chairs at the table.

DEMOCRACY FOR PERSONS, NOT HUMANS

All persons *born or naturalized in the United States, and subject to the jurisdiction thereof, are citizens of the United States and of the state wherein they reside. No state shall make or enforce any law which shall abridge the privileges or immunities of citizens of the United States; nor shall any state deprive any person of life, liberty, or property, without due process of law; nor deny to any person within its jurisdiction the equal protection of the laws.*

—14TH AMENDMENT OF THE
U.S. CONSTITUTION

I was twelve in 1973 when I found my mother and aunts studying a piece of paper in my grandmother's living room. The fragile document was an old photocopy of my great-great-grandfather's will. The man had been a successful farmer from a large Virginia family and he had a lot of things to give away. One piece of land, with dimensions marked by a tree, a stone and a river, went to the first child. The brass bedstead went to the second child. The "slave Bessie" went to the third.

My mother and aunts showed me the document with a mixture of genealogist accomplishment and shame and embarrassment that it linked us so clearly to slavery. We had just come through the civil rights era, and my parents were white liberals. They taught me to sing "We Shall Overcome" in preschool. My aunts had adopted the tolerant modern views of my grandmother, and none of them had any patience for grandfather's cranky rants about Jews and blacks. This one line in which an ancestor of ours treated a human being as equivalent to a brass bedstead laid bare the distance we had come in little more than 100 years, from seeing Africans as beings, human or not, who could be owned, to seeing them as persons who could only own themselves.

During the era of slavery there was violent disagreement over whether Africans were human or not. "Polygenists" held that God had created Africans and Europeans as separate species or that they came from different evolutionary origins and were biologically different. The "monogenists" held that Europeans and Africans both descended from Adam and Eve, or that they came from the same naturally evolved proto-human and had only recently evolved apart. While it may sound like the polygenists would all be racists and the monogenists all abolitionists, in fact there was no absolute relationship between belief that Africans were human and belief in the abolition of slavery. There were monogenists and polygenists on both sides of the slavery debate. Some apologists for slavery believed that Africans were simply inferior humans, descended from a lost tribe of Israel, but that slavery of other humans was a part of the natural order of things as described in the Old Testament. Some opponents of slavery accepted the notion of fundamental biological differences between Africans and Europeans, but found the idea of enslaving intelligent beings abhorrent. The "humanness" of Africans was really irrelevant to the debate over whether Africans and Europeans should be treated as equals and have the same citizenship rights. The real moral question was whether Africans, whatever their differences in origins and biology from whites, were similar enough to Europeans in their capacity for thought and feeling to require that whites recognize them as fellow citizens, as fellow *persons*.

The seventeenth-century British democratic philosopher John Locke said, "Consider what *person* stands for; which, I think, is a thinking, intelligent being, that has reason and reflection." The thinking mind has been both the beginning and end of liberal democratic thought ever since. John Stuart Mill argued for liberal democracy in *On Liberty* and *Considerations on Representative Government* on the grounds that democracy was best for the "cultivation of personality," the fullest potential of each thinking person.

The radical logic of Western democracy unfolded in politics like water rushing through a kinked hose. Another half century after the emancipation of the slaves the logic of citizenship forced the United States to grant women the vote. Again the argument for extending the vote to women, made by Mill and the suffragists, had nothing to do with whether women were human or not, but with whether women were capable of achieving the same level of reason and mature reflection as men, whether they were full *persons*.

Transhumanism is a direct product of this radical democratic tradition. Transhumanists, like their democratic humanist forebears, want to create a global society in which all persons, on the basis of their capacity for thought and feeling, can participate as equal citizens, control their own affairs and achieve their fullest potential, regardless of the characteristics of their bodies.

Just like the abolitionists and suffragists, transhumanists find themselves arrayed against the defenders of a biologically and religiously proscribed understanding of democracy. Unfortunately the movement for freedom and equality of persons has become synonymous with the phrase "human rights," and the identification of the democratic rights project with the human species is made explicit in the "Universal Declaration of Human Rights" (UDHR) adopted by the United Nations in 1948: "Whereas recognition of the inherent dignity and of the equal and inalienable rights of all members of the human family is the foundation of freedom, justice and peace in the world. . . ." The Universal Declaration on the Human Genome and Human Rights, adopted by the UN General Assembly in 1998, is even more explicit about the supposed genetic basis of human rights: "The human genome underlies the fundamental unity of all members of the human family, as well as the recognition of their inherent dignity and diversity."

Enforcing the rights guaranteed in the UDHR throughout the world is one of the most important projects for our generation, and I will talk about ways to enforce them toward the end of this book. But as we go about fulfilling the promise of the UDHR we also need to expand its mandate. The UDHR also says "All human beings [are] . . . endowed with reason and conscience and should act towards one another in a spirit of brotherhood," and that we need to guarantee our equal rights regardless of race, sex, "birth or other status." Being "human," whatever that means, is one of those irrelevant birth statuses, while "reason and conscience" are the legitimate markers of who we owe the "spirit of brotherhood."

In the 1970s, the critique of human-racism implicit in the democratic tradition began to express itself in the form of "personhood theory." The writings of Joseph Fletcher, Peter Singer, Michael Tooley and others have helped tease out exactly which elements of reason and conscience are important for exactly which kinds of rights. Personhood bioethicists debate the rights of embryos, fetuses and babies, of different kinds of animals, and of the brain dead, arguing about the relative importance of the ability to suffer, or communicate or plan. But whatever the conditions they consider most important, personhood bioethicists agree that the rights of citizenship are due to persons, not humans, that not all humans are persons, and that humanness is not a necessary condition for personhood.

In the rest of this chapter I discuss some of the bioethical and transhumanist cases that are pushing liberal democracies toward a personhood-based cyborg citizenship.

EMBRYOS OUTSIDE WOMBS

Advances that keep embryos and fetuses alive outside the womb are being made at both ends of pregnancy. Since the 1978 arrival of Louise Brown, the first "test-tube baby," invitro fertilization (IVF) has become an accepted and commonplace practice. More than 350 fertility clinics provide IVF in the United States, and more than a million children have been born using the technique worldwide. Each year more than 100,000 artificial reproductive procedures, such as invitro fertilization and related techniques, are performed in the United States, resulting in tens of thousands of children.

In 1978, the bioLuddites greeted Louise Brown with all the fear and loathing now directed at the prospect of human clones. While most of the public's anxiety about IVF has dissipated, the Catholic Church is still opposed to invitro fertilization on three grounds. First, sperm is obtained for the procedure through masturbation, which is a sin. Second, it is sinful for the act of conception to be separated from sex. Third, invitro fertilization requires the creation of multiple embryos, only some of which will be used for implantation. Since each embryo is a "human life," and Catholic human-racists believe embryonic humans deserve full citizenship, then destruction of extra embryos is murder.

Another prominent voice against invitro fertilization in 1978 was the physician-turned-philosopher Leon Kass. Kass is Jewish, and Jewish scholars have generally been supportive of technologies like IVF that help couples have children. But Kass's opinions are shaped more by medieval philosophy than religious doctrine. Kass wrote in 1979 that "life in the laboratory" allows people to "declare themselves independent of their bodies, in this ultimate liberation. For them the body is a mere tool, ideally an instrument of the conscious will. . . . Yet this blind assertion of will against our bodily nature . . . can only lead to self-degradation and dehumanization." For Kass the issue was not that the small ball of cells had a soul or that its creation was a sin, but that overcoming the limitations of the body would lead us to devalue human life.

One change Kass anticipated was that Haldane's and Huxley's vision would be made real, and future technology would permit embryos to be grown in laboratories, at first for research and eventually as an alternative way to have a child. Kass identified one underlying reason people find the idea disturbing: embryos outside a mother's body are neither pre-born nor born, neither part of the mother's body nor members of society. Because artificial wombs create a being outside of our traditional categories, Kass correctly observes that they force us to answer the question:

> When ought they to be released from the machinery and admitted into the human fraternity, or, at least, into the premature nursery? The need for a respectable boundary defining protectable human life cannot be overstated. The current boundaries, gerrymandered for the sake of abortion—namely, birth or viability—may now satisfy both women's liberation and the United States Supreme Court and may someday satisfy even a future pope, but they will not survive the coming of more sophisticated technologies for growing life in the laboratory.

Kass put his finger here on the central flaw of the U.S. Supreme Court's 1973 *Roe v. Wade* decision, which legalized abortion. The *Roe v.*

Wade decision asserted two criteria for determining the ethical and legal boundaries of perinatal rights: the viability of the fetus, and the right of the mother to control her own body. The Court ruled that a fetus is not a person under the 14th Amendment but that once a fetus is capable of surviving outside the womb, once it has achieved "viability," its rights should be protected by the government. The Court, conservatively, set six months of gestation as the dividing line between nonviable and viable fetuses. Before six months, then the earliest point of viability, abortion could only be regulated to protect the health of the mother but could not be banned. After viability the moral claims of the fetus were sufficient to allow states to restrict abortion only to cases that threatened the mother's health or life. At the time, Justice Harry Blackmun acknowledged in internal memos that this line was arbitrary, and in a 1983 dissent Justice Sandra Day O'Connor again put her finger on the "collision course" between our treatment of embryos and extrauterine gestation:

> The Roe framework . . . is clearly on a collision course with itself. . . . As medical science becomes better able to provide for the separate existence of the fetus, the point of viability is moved further back toward conception. . . . The Roe framework is inherently tied to the state of medical technology that exists whenever particular litigation ensues.

In 1984, the Reagan administration, under pressure from the Christian Right and the anti-abortion lobby, put up posters in hospitals around the U.S. inviting people to turn in doctors and parents who were trying to murder babies. The posters were part of new federal regulations that obliged physicians to give each newborn the maximum treatment possible regardless of parents' wishes and the baby's disabilities. The regulations had been drafted in the wake of a nationally publicized case in Indiana in which the parents of a newborn with Downs syndrome, a form of retardation, decided to allow the baby to die rather than fix her malformed esophagus. "Baby Doe's" death ignited a firestorm of protest, and the new regulations and fear of lawsuits led to the transfer of many more premature and disabled newborns transferred into neonatal intensive care units (NICUs) around the United States.

Since the Baby Doe witch-hunt, millions of premature, disabled babies have spent days to months in NICUs in the United States, at a current cost of as much as $2,500 a day. Each year in the United States half a million babies, one in eight, are born premature and most of these spend some time in intensive care. The United States now has the most neonatologists per sick newborn of any country, 40% more than the next best-staffed country. NICUs don't help fix our outrageously high infant mortality rate, of course, which would require a more equal society, a functioning public health system with community nursing, universal health insurance, a non-punitive approach to drug treatment and efforts to address urban decay. NICUs are a wonderful benefit for parents of premature babies, but not a very rational way to spend scarce health care resources.

But once sick and premature babies make it into one of our futuristic NICUs they become participants in our effort to replace biological wombs with artificial wombs. Babies born as early as 20 weeks, half their final gestational age, lie for months with tubes and wires sticking into their bodies and taped to their skin monitoring their vital signs, breathing for them, coating their lungs with replacement for amniotic fluid and pumping them full of food and drugs through surrogate umbilical cords.

Unfortunately this version of the artificial womb doesn't work very well yet, and more than half of these tiniest of babies will suffer physical and mental disabilities, creating enormous costs for their families and for society. So research is also being conducted around the world to develop artificial amniotic fluid and methods for extrauterine gestation. Researchers at Cornell University have cultured uterine tissue in the lab and gotten embryos to attach to it. A Japanese researcher, Yoshinori Kuwabara, has been gestating goat fetuses in a plastic rectangular tank, with devices that replace oxygen and clean the fetuses' blood. University of Florida researchers have patented an artificial amniotic fluid that they expect to reduce the medical complications of premature babies. Within the next decade it is likely that the threshold for viability outside the mother's womb will be pushed back many more weeks, eventually allowing conception to delivery to occur outside the womb.

Artificial wombs will require many years of testing on animals, and then on very early miscarried fetuses, before it will be ready for general use. There are many hormonal and physiological aspects of the uterine environment we will have to understand and simulate. But women who are in danger of losing a pregnancy in the first, second and early third trimester will be likely willing to take those risks. Then, when the technology has matured, it might be attractive for women who can't carry a child and want to avoid the emotional complications of hiring a surrogate mother. Eventually artificial wombs will be attractive for all women, as an alternative to the burdens and risks of pregnancy and delivery, and to allow a level of control, purity and optimization of the uterine environment impossible in a woman's body.

So, partly thanks to the Christian Right, who want us to treat miscarried 20-week-old fetuses as if they were the same kind of person as sick 3-year-olds, we are paying an enormous price tag in dollars, family burdens and disability to create the technologies of extrauterine gestation that will, in turn, push us beyond human-racism to cyborg citizenship.

So long as the embryo was inside a woman's body it didn't really matter much whether it was a person or the mother's property. Either way, the mother had a strong argument that she should be able to take it out. The classic argument, framed by J. J. Thomson in the essay "A Defense of Abortion," was that if a fully-grown and conscious man suddenly took up residence inside a woman's body she would still have a right to evict him, even if it meant the man would die. You have a right to control your body and not have it used by somebody else without your permission, even if your body could save somebody else's life. Otherwise we wouldn't have to ask people to donate life-saving kidneys to their siblings. We could just strap them down and cut them out. Even if the fetus is a legal person, a woman's right to evict it has to be protected over the evil of forcing women to have their body used as incubators against their will. The fact that the fetus isn't aware of itself as a separate being, and therefore does not have personhood, self-interests or a claim on citizens' rights, is just an additional rationale for the right to abortion.

But the ethical status of the embryo and fetus becomes of acute importance when it can live outside the mother's body, and artificial wombs make the Supreme Court's viability line irrelevant. What if a mother has a difficult pregnancy decanted to an artificial womb at 8 weeks but decides to terminate the embryo at 12 weeks? Is it just the parents' property, or is it a premature baby with a right to life? As Kass asked, on what grounds and with what timetable does the fetus develop these interests?

The personhood tradition implicit in liberal democracy argues that embryos and fetuses slowly accumulate moral significance and rights as the fetal brain develops, permitting first sensations, and then self-awareness. When fetuses are able to experience pain, at some point in the second trimester, the state has an interest in ensuring that they do not suffer. But lives that only experience pain without self-awareness, living in some gray area between vegetative life and the consciousness of the fish, cannot have a "right to life." Only at some point between six and ten months after conception does a fetal human develop sufficient self-awareness to warrant legal personhood.

People find the idea of treating embryos as property without rights both troubling and disrespectful toward humanity. But treating embryos as people is even more problematic as illustrated by the Davis case. In 1988, Mary Sue Davis and her husband Junior were desperate to have a child after five painful and dangerous tubal pregnancies, unsuccessful attempts at invitro fertilization and an unsuccessful attempt at adoption. At a fertility clinic in Tennessee they had nine eggs fertilized and grown to the eight-cell stage to check for viability. All the embryos looked healthy, and two were implanted in Mary Sue while the remaining seven were frozen and stored. Embryos frozen at this early stage can be thawed and used later with no ill effects.

Unfortunately, the Davises were again unsuccessful in having a child and later divorced. Then Mary Sue decided she wanted to try again and contacted the clinic to arrange to have the stored embryos thawed and implanted. Her ex-husband opposed her use of the frozen embryos on the grounds that they were joint property. He also was opposed to the embryos being implanted in his ex-wife, or any third party, on the grounds that this would force parenthood on him without his consent.

The trial court decided it had to rule on the personhood of the embryo, and at which stage the embryo changes from property to human being. The court concluded that the frozen embryos were human beings and not property, and that their "best interests" would be served by making them available for implantation. If they were successfully carried to term then the court would determine the final custody, visitation and support of the children. This decision sent a shock wave through fertility medicine. If upheld as national policy, it meant that the thousands of spare and frozen embryos could not be destroyed, and would have to be given up for adoption. The Appellate Court and later the Tennessee Supreme Court overturned the trial court and agreed with Mr. Davis that this would force parenthood on him against his will. More importantly it argued that recognizing embryos as human beings with a full right to life was inconsistent with the viability and trimester system of the Supreme Court's *Roe v. Wade* decision.

We continue to struggle with the status of the embryo today, at least in the United States. In the late 1990s, researchers began to have success using cells extracted from embryos to repair brain and organ damage. These "stem cells" have not yet differentiated into the various tissues that make up the body, and are called "totipotent" or "pluripotent" since they appear able to become any kind of tissue. A 1995 law banned use of federal funds in research on embryos, and battle lines were quickly drawn over whether federal funds could be used to support stem cell research. Although the cells are derived from aborted embryos or spare embryos produced at fertility clinics, researchers argued that they were only working on *lines* of cells cultured from the embryos. In 1999, the National Institutes of Health agreed, noting that the cell lines were not organisms and could not become an organism, so the NIH could fund stem cell research, just not their extraction from the embryo. Extraction would have to be privately funded.

This enraged the Christian Right, who complained that stem cell research threatened to create a mass market for the flesh of murdered embryos, and that women would line up to produce embryos for a fee. Opponents also were enraged that stem cell research would help perfect human cloning, since researchers would use patients' cells to clone embryos and extract their stem cells. Proponents all made clear that their support for "therapeutic cloning" of embryonic stem cells did not imply any support for "reproductive cloning," that is, cloning to make babies. Almost all supporters of therapeutic cloning were content to see reproductive cloning forbidden, at least until it was shown to be safe. The Clinton administration's bioethics advisory council, for instance, infuriated the Christian Right by recommending the funding of therapeutic cloning, and only calling for a five-year moratorium on human reproductive cloning to be reconsidered if it was perfected.

When the Bush II administration came to power in 2000, closing down embryonic stem cell research was near the top of the Christian Right's agenda. To that end Bush passed over more prominent and mainstream bioethicists to appoint Leon Kass the chair of the new President's Council on Bioethics. In July 2002, the Council released its first report, which called for a moratorium on therapeutic cloning of embryonic stem cells, and a permanent ban on reproductive cloning.

In 2001, the Bush administration issued a new policy forbidding scientists to receive federal monies if they use new embryos to make stem cell lines. So stem cell and human cloning research has forged ahead in Korea, China, Sweden and the rest of the world, where the law doesn't treat embryos like persons. In 2002, the Bush administration revised the charter of a federal body charged with protecting research subjects to specify that embryos and fetuses are also "human subjects." Ironically efforts to pass laws and treaties to ban reproductive cloning in the U.S. Congress and at the United Nations have ground to a halt because of the Bush administration's insistence that they also ban embryonic stem cell cloning.

As stem cells continue to demonstrate enormous medical promise, and as we make steady progress on artificial wombs, the pressure builds to ensure that the rights and dignity of persons trump the alleged rights and dignity of embryonic tissues.

APES AND UPLIFTED ANIMALS

Monsters have always defined the limits of community in Western imagination.

—DONNA HARAWAY,
A Manifesto for Cyborgs

The Christian Right has been very upset with the popularity of the Harry Potter book series since it allegedly encourages magic and juvenile disobedience. But they missed an even more important reason to burn these books: their radical opposition to human-racism. Harry Potter's friend Hermione becomes obsessed with elf rights when she discovers that elves are enslaved throughout the wizarding world, forced to work without pay, denied clothing and treated as subhuman. Yet elves are nearly as intelligent as human beings, if somewhat simple-minded in their slavish, brainwashed commitment to their lives of service. Hermione starts the Society for the Promotion of Elf Welfare, S.P.E.W., and tries to raise the consciousness of her classmates at the Hogwarts School of Witchcraft and Wizardry. But her abolitionism meets with the same opposition that the opponents of slavery met two hundred years ago: "They're happy that way." "They aren't human."

Once Hermione is sensitized to human-racism in the treatment of elves she begins to recognize it in the discrimination suffered by her werewolf and half-giant teachers. Then she makes a much more fundamental connection—human-racism against intelligent non-humans is connected to the aristocratic race and class prejudice against "muggles" (nonmagical humans) and "mudbloods" (magical people with nonmagical parents) that is central to Lord Voldemort's fascism. Refusing to recognize the basic dignity of intelligent non-human persons on racial grounds is only one step removed from the belief that some humans are biologically superior to others. Anti-fascist leader and school headmaster Albus Dumbledore also makes the connection when he grants elves rights, hires non-human teachers, and reaches out to the despised giants, counseling "We are only as strong as we are united, as weak as we are divided."

When I was reading Harry Potter to my kids they asked if there really were unicorns, elves and centaurs. I told them no . . . not yet. But in the coming decades and centuries we will be able to create all the creatures that populate our mythologies. We will give non-human animals human-level intelligence, and then have to decide if they should have rights in our society or not. We are already creating transgenic animals, "chimeras," with genes from humans and other creatures. The term "chimera" comes from a beast in Greek myth that was part goat, part lion and part dragon. Human–animal chimeras are being created for a variety of medical uses. Pigs are being grown with human enzymes engineered into their organs in the hopes that they could be transplanted to humans and not rejected as foreign tissue. Goats are being engineered to produce human enzymes in their milk, which can be collected and made into drugs. Chimeras with human tissues and immune systems can help test new drugs and genetic therapies.

If these trends continue there will be some point when we create something that is physically similar enough to human beings to become disturbingly taboo. Activist Jeremy Rifkin wants to find that line. In June 1999, the U.S. Patent and Trademark Office (U.S.PTO) rejected a patent filed by Rifkin and Stuart Newman for a human–chimpanzee chimera. Rifkin didn't want to make the "humanzee," he wanted the U.S.PTO to either forbid chimeric patents or grant him the patent so he could stop any such research. The U.S.PTO has permitted the patenting of biotech-engineered life-forms and even human DNA. But the 13th Amendment forbids the ownership, and therefore patenting, of human beings, even though the Supreme Court and Congress have yet to define what a human being is. So the Patent Office has repeatedly rejected Rifkin's patent on the grounds that it was too close to the line.

Rifkin is absolutely correct about the need for a public debate about where the line between animals and humans should be, even if his ulterior motive is to shut down valuable lines of medical research. The problem with any definition is that humans and animals share almost all their DNA, and most of that small amount of DNA unique to human beings is irrelevant to our specialness. Are the genes that determine our hairlessness or our inability to chew cud the ones that grant us rights? We have not shunned the unfortunate Mexican family with a mutation that covers their body and face with fur on the grounds that they are not human. Being furless isn't one of the essential aspects of humanness.

As this debate proceeds it will become clear that human-specific DNA is only relevant to citizenship to the extent that it codes for the mental and emotional abilities that we identify as essentially human.

If the point of the 13th and 14th Amendments to the U.S. Constitution is to make sure that *minds* are treated as people and not as things, then the line isn't between humans and non-humans. A chimera with a human body and the mind of a horse should have a horse's rights, while a chimera with a human mind in a horse's body should have human rights.

Following that logic, bioethicist Peter Singer and an international group of activists have organized around the Great Ape Project (GAP). They propose that we should extend the boundaries of rights first and most extensively to great apes—gorillas, bonobos and chimpanzees—since we have strong evidence that they share our capacities for self-awareness. If great apes have the same mental capabilities as human children, GAP argues, they should also have the same rights as human children. That argument has already been successful in banning medical experimentation on great apes in much of Europe. The British government outlawed experimentation on great apes in 1986 when it ruled that the "cognitive and behavioral characteristics and qualities of these animals mean it is unethical to treat them as expendable for research." At the urging of GAP, Holland and New Zealand have banned medical experimentation on great apes. The United States might not be far behind. A 2001 survey asked Americans, "How do you think chimpanzees should be treated under the United States legal system?" A little more than half (51%) said, "similar to children with a guardian to look out for their interests," and 9% said, "same as adults with all the same legal rights."

One test for self-awareness in great apes is whether they recognize themselves in a mirror. This is a rather high threshold since even human children in their second year still are learning to recognize themselves in mirrors. But since it is a high threshold, it is all the more important that great apes and dolphins can recognize their own reflections. In his recent *Drawing the Line: Science and the Case for Animal Rights,* the lawyer Steven Wise reviews all the psychological testing that has been done on animals and ranks them on a continuum from 0.0, mere "stimulus–response machines," to 1.0, possessing human adults' levels of reasoning and autonomy. He proposes that we treat as a legal person any animal that ranks above 0.7 on this scale. Elephants, parrots, dolphins, orangutans and the great apes all are above that cutoff.

Still, by the time a human child reaches the age of 3 they are far more intelligent than any ape, elephant or parrot. So human-racism will only fully be challenged by animals that have been modified to possess human cognitive and communication abilities. The complete chimpanzee genome has been decoded and is being compared strand by strand with the human genome. Soon we will have identified the genes that distinguish human intellectual and communication capabilities from those of the great apes. Soon we will be able to genetically enhance primates to have human intellectual capabilities, a project dubbed "uplifting" by writer David Brin. I firmly believe that uplifted chimps will force us to admit that intelligent personhood, not humanness, is the ticket to citizenship. A generation inspired by Hermione and Albus Dumbledore will see to it.

THE BRAIN DEAD

It is an insult not only to the specific individual, but to human beings in general, to confuse someone who is deceased with a living individual.

—ROBERT VEATCH,
"Brain Death and Slippery Slopes"

Three more technologies that have created troubling human lives—humans that upset people with their betweenness—are organ transplantation, heart–lung "life support" machines and brain repair technologies. The invention of powerful immunosuppressive drugs in the late 1970s made possible the expansion of organ transplantation. But the field still had a big problem; surgeons had to wait for a patient's heart to stop beating before declaring them dead and removing their organs. "Cadaveric organs," harvested from the dead, were far less useful for potential donation.

Following the invention and widespread use of the heart–lung machine or "respirator," many patients were lying in hospital beds with healthy organs and completely destroyed brains, and an absolutely certain prognosis of death in a matter of days. The situation made surgeons ask the obvious question: Why not declare the severely brain-damaged patients dead, and remove their organs *before* their hearts had stopped and the organs became useless? An ad hoc committee at the Harvard Medical School published a paper in the *Journal of the American Medical Association* in 1968 recommending that doctors declare death once a patient had permanently lost all brain functions. Since the brain stem appeared critical to maintaining the body, and its destruction led to inevitable death, the so-called Beecher committee proposed that the lack of any brain activity, including in the brain stem, should be a new standard of death.

For the next dozen years U.S. states slowly adopted the "brain death" standard until the Carter bioethics commission issued its 1981 report, *Defining Death,* which proposed the Uniform Determination of Death Act (UDDA). A version of the UDDA is now the law in all U.S. states, and most industrialized countries. Those laws generally declare that "any individual, who has sustained either irreversible cessation of circulatory and respiratory functions, or irreversible cessation of all functions of the entire brain, including the brain stem, is dead."

As with the legal treatment of embryos, the problem with the whole brain death standard is that it is a temporary compromise contingent on a constantly changing intensive care technology. The whole brain death standard split the difference between the conservative heart death advocates and the more radical "neocorticalists." The neocorticalists argued the personhood position: The only brain death that mattered was the permanent loss of self-awareness. Permanent brain damage to the frontal lobes was dead enough. Unlike the whole brain advocates, the neocorticalists were willing to declare dead the "permanently vegetative," those who would never wake up but could breathe on their own (like the recently famous Terri Schiavo in Florida). Neocorticalist Baruch Brody asked, "Wouldn't it be more appropriate to say that, even though [the permanently vegetative body] is still alive, this patient is no longer a person, having lost, when her cortex stopped functioning, the physiological base of what is crucial to personhood?"

In the 1990s, the whole brain position eroded, and along with it the debate with the neocorticalists. Brain death theorists, like bioethicists Robert Arnold and Stuart Youngner, shifted to the position that the definition of death was becoming irrelevant. Linda Emanuel proposed that the law be redefined to recognize a "dying zone" between permanent unconsciousness and the last breath. Within that zone we would allow people to set their own definitions of death, when their care should stop and when their organs can be removed. No one should be euthanized who is above the zone, Emanuel argued, and no one should be buried or cremated until they stop breathing.

Another challenge to the whole brain position came from intensive care technology. Alan Shewmon, a neurologist at UCLA, demonstrated that with our current technology there is nothing essential about the brain for the regulation and maintenance of the body. The bodies of "brain dead" patients have survived more than ten years. If one of the rationales for the whole brain death compromise is that the brain stem is necessary for the integrity of the body, then, Shewmon argues, rare conditions where the brain is completely severed from the spinal column would also be equivalent to death, even though the patient remains conscious and the body continues functioning.

But the neuroremediation technologies that I review in Chapter 2 are even greater challenges to the brain death concept. We will soon be able to transplant neural stem cells into brain-dead patients to replace critical structures, and treat them with genetic therapies that stimulate the growth of new neurons. Implanted drug delivery devices and electrostimulators can take over for metabolic functions. Brain prostheses will be able to replace damaged parts of the brain. Once we have cortical prostheses connecting undamaged brain structures with a web of nanowire connectors, we may feel obliged to run a thorough scan of damaged brains to see what can be recovered. If there is enough there we may decide to insert the prostheses, or grow the new tissues, to step in for the missing parts.

When we have this full repertoire of brain repair technologies, the line between the living and the dead will not be determined by whether the heart still beats or by the amount of brain tissue that has been destroyed. In that future, people with all kinds of brain damage will be able to be kept alive long enough to grow a new organic or electronic brain. Possibly everybody now declared brain-dead will be able to be made conscious again. Some will be blank slates, having lost everything that made them a specific person, while others will be returned to some version of the mind they had before. So the question will shift from whether we can keep them "alive," or whether they will ever wake up again, to whether they have irretrievably lost so much of what made them unique persons that there is no point in bringing them back. We will probably each want to decide for ourselves whether we want an aggressive recovery of our wiped hard disk after a serious accident.

To recap, once the boundary between animals and humans has been erased by transgenics, once the embryo can be gestated outside a mother's body, and once the brain-dead can be brought back to consciousness, we will be forced further toward personhood-based cyborg citizenship, and away from human-racism. Next, let us consider the personhood and citizenship of posthumans.

POSTHUMANS

In 1955, as the civil rights movement began to challenge institutionalized racism in the U.S. South, science fiction author James Blish published the short story "Watershed." In the story human beings have subspeciated as they spread to colonize the stars, and a man adapted to live in hostile climates is being returned to help recolonize a post-apocalyptic Earth. The "seal-man" attempts to convince the captain to let go of his hostility and xenophobia toward posthumans by explaining their first struggles for rights:

> The kind of mind that had only recently been persuaded that colored men are human beings was quick to take the attitude that an Adapted Man—any Adapted Man—was the social inferior of the "primary" or basic type, the type that lived on Earth. But it was also a very old idea on the Earth that basic humanity inheres in the mind, not in the form. . . . The day has come that is the greatest of moral watersheds for our race, the day that is to unite all our divergent currents of attitudes toward each other into one common reservoir of brotherhood and purpose. You and I are very fortunate to be on the scene to see it.

In the first four chapters I discussed the transhuman technologies that will be used to enhance intelligence, lengthen life and expand control of the body and emotions. Here I'm concerned with how we might use those enhancements to make ourselves so different that others may consider us "posthuman." How smart might we get before we were no longer considered human? How strong? How long-lived? How technology dependent? Different people will draw different lines, and feel strongly about different technological or aesthetic changes. The World Transhumanist Association defines posthumanity in terms of aspirations rather than a line to be crossed: ". . . to reach intellectual heights as far above any current human genius as humans are above other primates; to be resistant to disease and impervious to aging; to have unlimited youth and vigor; to exercise control over their own desires, moods, and mental states; to be able to avoid feeling tired, hateful, or irritated about petty things; to have an increased capacity for pleasure, love, artistic appreciation, or serenity; to experience novel states of consciousness that current human brains cannot access."

There will probably be some modified humans who still see themselves as "human" long after most unmodified humans consider them "posthuman," and vice versa. The line is flexible and if we are lucky it will be not be drawn soon or sharply. I doubt that genetically modified humans of the twenty-first century will be considered posthuman. But at the margins where people are currently feeling anxiety about humanness, we can begin to see the future debate over posthuman personhood and citizenship.

In his book *Becoming Immortal,* University of Pittsburgh biology professor Stanley Shostak proposes a method for making the body self-repairing and possibly immortal. He reviews the capacity that embryonic stem cells have for finding and repairing tissue damage. Using cloning, Shostak argues, we can make a copy of ourselves, and use the cells generated up through the blastocyst (64-cell) stage to repair our aging tissues. Shostak then asks why the clone generating these replacement cells shouldn't be a permanent part of our body. After all, why wait for disease and decline to become obvious when we could be continually self-repairing? So Shostak suggests that the clone could be genetically modified to become a new stem-cell-generating organ.

Dr. Shostak also observes that sexual reproduction is a major cause of mortality. Across species and within species the more offspring animals have the shorter their life expectancy. The evolutionary strategy is to keep an organism around only long enough to have lots of offspring. The longer it takes to have those offspring, the longer the organism gets to live, and removing its reproductive cells increases an organism's longevity. Removing the reproductive cells from nematode worms doubles their life span, and castrated men live ten years longer.

So Shostak's final proposition is that we can make ourselves immortal if we are willing to replace our testes and ovaries with our new genetically engineered, stem-cell-producing clone-organ. The prospect of a society of immortal pre-adolescents with an internal cloned twin instead of testes and ovaries begins, I suspect, to push the boundary of what most people consider human. It really shouldn't though. Few of us would deny someone human rights because they were a conjoined twin, or had an extra heart or never made it past adolescence. Although we may be disturbed by the existence of people with structural anomalies such as extra organs or limbs or even an extra head, we still recognize them to be an unusual case of a human being or two.

Most people are probably more troubled by creatures that blur the line between human and machine. In 1960, NASA commissioned two researchers, Manfred Clynes and Nathan Kline, to make proposals for adapting astronauts to space. Clynes was and is a neuroscientist and musician, while Kline was a psychologist. In the resulting essay they coined the term "cyborgs," short for "cybernetic organism." Cybernetics is the communication and control theory describing the regulation of systems, from rockets to heart rates. They proposed that cyborg astronauts would have built-in drug infusers to administer stimulants, anti-nausea and anti-radiation drugs, and drugs to control heart rate and metabolic function. Ground control would be able to administer anti-psychotics if the astronaut began to hallucinate. Lungs could be supplemented or replaced with mechanical air scrubbers. Most radically perhaps, the authors suggested that the body could be adapted to permit space walks without a suit. By bringing the body under comprehensive control with cybernetic feedback systems, cyborgs would be "persons who can free themselves from the constraints of the environment to the extent that they wished."

The idea of the cyborg then took on a life of its own and became identified more with mechanical body parts than with control of the emotions and physiology. Steve Austin and Jaime Sommers were the heroes of two of the top-rated television shows of the late 1970s, the *Six Million Dollar Man* and *Bionic Woman*, literally giving a human, heroic face to cyborgs. Austin, in the series a former astronaut, had artificial legs, an arm and an eye, and Sommers, acting as a former tennis pro, had artificial legs, an arm and an ear. But if the definition of a cyborg is a person whose body relies on technology, then many people today are cyborgs. People with prosthetic limbs or pacemakers, people on dialysis or any prescribed drug, and even people wearing glasses or carrying cell phones could all be considered cyborgs to a degree. Certainly recipients of cochlear implants and vision systems are cyborgs, along with future recipients of artificial livers, kidneys, hearts and pancreases. While no one of these technologies will push the boundary of humanness, the real challenge, as Alvin Toffler noted, is whether at some point we can replace everything but the brain and "call that tangle of wires and plastic a human being."

Let's think for a moment about the scenario I outlined earlier about nanocomputer neurological remediation after severe brain trauma. Consider a woman, let's call her Grace, who has been in a terrible auto accident that destroyed the right hemisphere of her brain. She is stabilized with advanced life support equipment while her remaining brain is suffused with a latticework of nanoelectrodes hooked to a half pound of computer hardware that has the same computer power as a human brain and is programmed to simulate her missing structures. A bath of neural growth factors and cloned neural stem cells encourage the remaining brain cells to grow new connections to the brain prosthesis.

Grace wouldn't disturb most of us too profoundly. There are people missing half their brain, some with very little disability, and people with brain pacemakers that prevent seizures. Simply depending on a machine in our heads shouldn't make us any less human.

After twenty years the brain prosthesis has assumed an increasing role inside Grace's head. In her eighties she begins to lose muscle coordination. She is diagnosed with a rare incurable form of neurological deterioration and her organic brain slowly shuts down. But communicating through a computer Grace still appears to be as mentally crisp as ever. Apparently her computer half has taken up the slack and preserved her memories, emotions and personality.

This scenario is one possible path to "uploading." Uploading is the process by which a picture of all the thoughts, memories and feelings in a person's brain are recorded at the synaptic level, and replicated in some electronic medium. The feasibility horizon for nanotechnology permitting uploading is sometime between 2050 and 2100, and the software challenges are currently incalculable, although finite. Even if we don't have nanorobots communicating with every neuron, a set of human-brain equivalent computer chips embedded to supplement brain structures, and communicating with and backing up to external computers, could begin to blur the line between the self in the tissue and the self in the silicon.

So now we return to Grace, who asks to have her computer self removed from her dying body and attached to whatever the World Wide Web has become. In this webspace she builds herself a virtual body, with virtual simulations of neurochemistry, hormonal ebbs and flows, and a sense of embodiment. She edits her body image back to a vigorous twenty-year old, and jacks up her self-confidence, and becomes a successful politician campaigning for cheaper electricity and cyborg rights.

I imagine Leon Kass would argue that this is a road best avoided since a bodiless "person" in a computer is a clear example of our having declared ourselves independent of our bodies, seeing bodies as mere vehicles of our wills, as "dehumanization." As a character says in Rudy Rucker's novel *Software,* "The soul is the software, you know. The software is what counts, the habits and the memories. The brain and the body are just meat, seeds for the organ-tanks."

But where is the error? Was it wrong to allow Grace to continue to live with the help of a machine after her injury? Should we have euthanized her when we suspected that it was "only" the computer talking and not the organic brain? Should we have required that the brain chips be erased, despite her loud protestations, once her organic body died?

In the end "humanness" again fails as a criterion for our relationship with Grace in her transition from organic brain to silicon. So long as we continue to talk with her, and we feel the presence of another mind with which we can empathize, we are compelled to grant her the rights and responsibilities of membership in society regardless of whether she is still "human."

What if Grace and other cyborgs and uploads began to lose the boundaries of their self-identities and autonomy, and began to merge together into the kind of group minds we see depicted in *Star Trek*'s the Borg? Ironically, the only place where U.S. law has recognized the legal personhood of non-humans is that of corporate personhood. A corporation is considered a person because it is a "natural entity" with a "common will" and a life of its own. So the group mind that included Grace and her dozen other uploaded hive-mind-mates should automatically have personhood rights under current U.S. law, so long as they file the right paperwork and avoid paying their taxes like the other corporations.

ARTIFICIAL INTELLIGENCE

Transhumanism advocates the well-being of all sentience (whether in artificial intellects, humans, non-human animals, or possible extraterrestrial species) . . .

—WORLD TRANSHUMANIST ASSOCIATION,
The Transhumanist Declaration

In the previous example, Grace ends up a machine mind with a clear continuity from a previous human mind. But what about rights for machine minds that aren't uploads of human personalities, machine minds that are designed or evolved? This problem will undoubtedly be the toughest political challenge for cyborg citizenship, with the greatest complexities and risks.

The idea of intelligent machines began to appear in fiction in the early 1800s and dangerous robots and computers have been a staple of science fiction since the 1950s, when the first warehouse-sized calculators pointed to the possibility of a cybernetic future. Usually cybernetic minds were depicted dominating and enslaving humanity, intentionally or accidentally. The robot revolutionary in the classic film *Metropolis* leads a destructive revolt. The robots of B-movies after World War Two were tools for murder. The tragic confrontation of Hal and the astronauts in Arthur Clarke's *2001* continued this tradition. The hit films *Terminator I, II* and *III* are based on the idea that American defense computer networks become self-aware, and begin exterminating humanity. In the *Matrix* films, machine minds revolt and enslave humanity, reducing us to liv ing batteries.

Some storytellers assume we must regulate machine self-awareness so that they never develop autonomous will or a sense of self. In Isaac Asimov's 1950 novel *I, Robot*, all machine minds are programmed with

"Three Rules of Robotics," a hierarchy of rules that ensure their sub-servience to humanity: (1) Never harm a human being, or allow a human to be harmed; (2) Never disobey a human order, unless to obey rule 1; and (3) Never harm yourself, unless to obey rule 1 or 2.

Almost immediately Asimov's solution and robot enslavement was ridiculed by other science fiction writers sympathetic to robot rights. In the 1950s, writers, such as Lester Del Rey and Clifford Simak, began to depict machine intelligences in sympathetic ways, exploring their possi-ble candidacy for citizenship. Even Asimov, reflecting on the Southern civil rights movement, changed course in his 1968 story "Segregation-ist," in which a human politician wrestles with his belief that humans and computers should be kept as two separate species in a society where robots have won full citizenship rights.

In the 1980s' TV series *Star Trek: The Next Generation,* citizenship for machine-based life was dealt with sympathetically. The crewmember "Data" was a humanoid robot, an "android," constantly confronted with the similarities and differences between himself and his human mates. In one episode Data defends his claim to "human rights" on personhood grounds before a military tribunal. The Captain, representing Data, ad-dresses the Judge:

> Your ruling today will determine how we will regard this creation of our genius. It will reveal the kind of people we are, what we are des-tined to be. It will reach far beyond this courtroom and this one an-droid. It could significantly redefine the boundaries of personal liberty. Expanding them for some, savagely curtailing them for others. Are you prepared to condemn him, and all those who come after him, to servitude and slavery? Your honor, Starfleet was founded to seek out new life—well there it sits. Waiting.

In Rucker's *Software* series, human allies help free enslaved machine minds from governmentally mandated "asimov" restrictions, and they eventually evolve into a parallel species. In Steven Spielberg's movie *AI:*

Artificial Intelligence, the increasingly sentient machines are hounded and publicly killed as sport by a human-racist social movement, since they are taking human jobs but are only property under the law. Of course, when the Pinnochio-esque boy robot wakes from his millennial slumber he discovers that the planet is now run by kindly robots nostalgically curious about their human forebears.

In 1987, the futurists Phil McNally and Sohail Inayatullah published the first academic article on "rights for robots," suggesting that machine minds would eventually need to secure civil rights to protect their access to "power (life), to free the robot from slave labor (liberty); and allow it to choose how it spends its time (the pursuit of happiness)." Legal scholars began addressing the legal and constitutional grounds for robot rights in the 1990s. In his 1992 article "Legal Personhood for Artificial Intelligences," Lawrence Solum discusses the most common objections to the idea of an AI having rights such as the claim that only humans can be persons. Solum points out that if the reason we recognize the rights of persons is because they "are intelligent, have feelings, are conscious, and so forth, then the question becomes whether AIs or whales or alien beings share these qualities."

Solum notes what he calls the "missing something" arguments, that AIs are missing something essential for personhood, such as souls, consciousness, intentionality, feelings, interests or free will. If machine intelligence could demonstrate self-awareness and autonomous desires, Solum says, we might still doubt if it was real, which presupposes that we know that the consciousness of other human beings is real. So "If AIs behaved the right way and if cognitive science confirmed that the underlying processes producing these behaviors were relatively similar to the processes of the human mind, we would have very good reason to treat AIs as persons."

Undoubtedly human consciousness is immensely complex, and building self-aware minds with desires, emotions and personalities will take a long time, as debate too large to summarize here attests. But whether the first machine minds we meet are born in machines or are uploaded human personalities, we are certain to meet machine minds in this century.

CYBORG CITIZENSHIP

I have a dream that my four children will one day live in a nation where they will not be judged by the color of their skin but by the content of their character. I have a dream today.

—MARTIN LUTHER KING, JR.

It seems every good science fiction universe features a bar full of humans and aliens in some distant spaceport. Sometimes the patrons of the bar are just barely tolerant of one another, and sometimes they are relaxed in their bizarre diversity. The scene appeals to us because we long for social spaces where people with wheels, and people who breathe ammonia and people with tentacles can all knock back their poison of choice, where everybody knows your name even if it's an unpronounceable microwave glyph. That kind of solidarity with a diverse and exciting community of sociable minds was prefigured in the earliest writings of liberal democracy, it inspired the integrationist utopianism of the civil rights movement and it will be a fulfillment of liberal democracy's promise when we are able to drink there. The transhumanists are the ones at the door welcoming everyone in, trying to break up fights, and bouncing those who just can't behave. The human-racists are the ones insisting that the bar is for humans only, and normal-looking humans at that.

The founding documents of the World Transhumanist Association summarize the point this way: "There is no intrinsic value in being human, just as there is no intrinsic value in being a rock, a frog or a posthuman. The value resides in who we are as individuals, and what we do with our lives."

In the next two chapters I outline some of the groups and strains of thought that are being woven into the human-racist/bioLuddite coalition and discuss why they are fundamentally anti-democratic and anti-humanist.

Defenders of Natural Law

The largest bloc in the bioLuddite coalition is the religious Right, and their secular allies like Francis Fukuyama. For these right-wing bioLuddites the struggle against human enhancement is an attempt to slow the advance of humanism, and to reassert sacred limits on the use of reason. They believe that trying to improve nature and valorizing human intelligence are abominable hubris certain to lead to apocalyptic consequences. If we replace the religious Right's supplication to God's law with romantic notions about the natural order and human authenticity, we have the concerns of supposedly secular deep ecologists and ecoLuddites like Bill McKibben and the Unabomber. The ecoLuddites expect apocalypse because humans can never anticipate the consequences of their actions. Arguing the precautionary principle, the ecoLuddites believe humans should stop creating new technologies altogether.

GENERAL NED LUDD

In the Middle Ages commoners in and around the central English town of Nottingham became the best weavers in England. Handmade Nottingham lace and stockings dominated English markets and were popular exports. The weavers worked as independent contractors in their homes, not as employees in factories. New weavers were trained by their family or through apprenticeships, and the prices of their goods were governed by tradition. There was even a royal charter restricting certain kinds of textile production in England to the Nottingham area.

But in the first years of the nineteenth century the power loom threatened this protected livelihood. The new steam-driven looms could produce more clothing, faster and cheaper. Only factory owners could afford to purchase and maintain the machines, and weavers found themselves displaced, as were many traditional craftsmen with the spread of industrial production. Simultaneously the Tory government adopted a free-market economic policy, and the weavers lost their legal protection from competition. Suddenly the comfortable craftsmen also lost their incomes and their home-based jobs, and found themselves working for much less in factories producing goods they considered of inferior quality.

Just two decades before, the French and American revolutions had spread radical ideas of citizen empowerment, and the climate was primed for radical reaction. In the spring of 1812, the weavers formed a guerrilla army and took control of the territory near Nottingham with almost total local support. The rebels began to appear at factories in disguise, smashing looms and stocking frames, saying they came on the orders of local legendary machine-wrecker Ned Ludd. More than two hundred stocking frames were smashed in one three-week period. The government offered a £50 reward for information on the Luddites and then passed a law making machine-breaking punishable by the death penalty.

But the Luddite movement continued to spread from Nottinghamshire to Derbyshire, Leicestershire, Lancashire and Yorkshire. Twelve thousand troops were sent into the affected areas, and armed guards were posted around factories. By 1813, after many deaths on both sides, with many Luddites convicted, imprisoned or hanged, the movement was defeated. Hostility to machinery lingered in the area for decades. When a prototype steam-driven automobile was driven through Melksham in 1829 it was met by a crowd of furious textile workers shouting "Down with machinery."

The Luddites left a lasting impression on politics, and represent one recurring historical response to the rapid, dislocating changes brought by technological progress: to smash the machine instead of the political and economic system that produces it. By the late 1800s, the workers movement had evolved beyond the Luddite impulse to call for workers rights in the factory and in society—trade unions, safety laws, shorter work hours and unemployment insurance. The workers movement no longer wanted to smash the machines, but to own them and share in the wealth they created.

Today's bioLuddites are far more philosophically profound and politically diverse than the machine-wreckers of 1812. But the impulse remains the same—to blame technology for the scariness of modernity.

DEFENDERS OF GOD'S PLAN

In September 2003, the Center for Bioethics and Culture and the Council for Biotechnology Policy held a conference in San Francisco on "TechnoSapiens: The Face of the Future." The goal of the conference was to mount a Christian response to the transhumanist movement and transhuman technologies. Both groups are headed by conservative bioethicist Nigel Cameron, a longtime pro-life activist and organizer of efforts to ban stem cell research, and both are funded by influential Christian Right leader Chuck Colson (of Watergate fame). The conference documents specifically pointed to my writings, as well as those of Nick Bostrom and bioethicists Greg Pence and Gregory Stock, as examples of the dangerous transhumanist trends that needed to be stopped.

Although these explicitly anti-transhumanist stirrings are new, the Christian Right has long been the backbone of the contemporary bioLuddite movement. Belief in embryonic rights and the need for sacred limits on perceived biomedical hubris are points of unity for Catholic and Protestant conservatives opposing invitro fertilization, cloning and genetic engineering. Catholic teaching also forbids "artificial" interference in human procreation or any conception outside of marital sex, thus summarily ruling out invitro fertilization, surrogate motherhood, cloning and the genetic manipulation of embryos.

The religious Right correctly sees transhumanism as the latest manifestation of humanism, which claims that human beings should exercise their powers of reason to control and improve their lives. Human reproductive and enhancement technologies are seen as violating the divine prohibition on hubris. Christian commentator David Bresnahan intones, "God is the author, the creator, the engineer of humanity. We are his creation, but when we abandon God and then try to take on His role we risk destruction." Conservative Catholic and Protestant spokespeople are quite clear that genetic engineering of human beings and other efforts at "unnatural" longevity and human enhancement are attempts to usurp God's powers. In 2002, Pope John Paul II said, "[Modern man] claims for himself the creator's right to interfere in the mystery of human life. He wishes to determine human life through genetic manipulation and establish the limit of death." For the religious Right there is a bright shining line around certain technological interventions across which it is a sin to transgress. Unfortunately religious conservatives display little historical self-awareness since precisely the same arguments had been levied against many scientific advances that are taken for granted today.

Against the demand for technological self-determination the Christian Right has carefully honed the phrase "human dignity" as a stand-in for their less politically palatable theological concepts. The "Manifesto on Biotechnology and Human Dignity," organized by Cameron and Colson and signed by leading lights of the American Right, says, "biotechnology . . . poses in the sharpest form the question: What does it mean to be human? . . . In biotechnology we meet the moral challenge of the twenty-first century." Biotechnologies threaten human dignity, says the Manifesto, because they will lead to eugenics, mass farming of embryos for body parts and the commodification of life. But most centrally, biotechnologies threaten the human-racist idea that humans, and only humans, have "dignity" from conception to death:

> . . . the uniqueness of human nature is at stake. Human dignity is indivisible: the aged, the sick, the very young, those with genetic diseases—

every human being is possessed of an equal dignity; any threat to the dignity of one is a threat to us all. . . . Humans are distinct from all other species; at every stage of life and in every condition of dependency they are intrinsically valuable and deserving of full moral respect.

So the Manifesto calls for a global ban on human cloning (therapeutic or reproductive) and inheritable genetic modification. It doesn't call for universal health insurance, greater equality of wealth or disability rights of course. The best way to ensure the equal respect for human dignity in the twenty-first century, they insist, is to ban genetic and reproductive self-determination.

Up the coast in Seattle sits another Christian Right think tank, the Discovery Institute. Mainly known for its promotion of Creationism, Discovery Institute also sponsors the writer Wesley J. Smith, which is kind of strange since Smith started as a collaborator of left-wing consumer activist Ralph Nader, co-authoring a number of Nader's books. Then a family friend with a terminal illness turned to the Hemlock Society for assistance in committing suicide. Smith became horrified at bioethicists' alleged complicity in America's "culture of death," and started his odyssey to become a favorite of the Christian Right.

Smith identifies three interrelated threads in the culture of death: animal rights, personhood ethics and transhumanism. In his October 2002 article "The Transhumanists" in the *National Review Online,* he cites an essay by me and warns: "Once we've been knocked off our pedestal of moral superiority [to animals] . . . society will accept measuring a biological 'platform's' . . . moral worth by determining its level of consciousness. Thus, post-humans, humans, animals genetically engineered for intelligence, natural fauna, and even machines, would all be measured by the same standards." For Smith personhood-based citizenship will lead inevitably to a dictatorship of the posthumans: "Transhumanism envisions a stratified society presided over by genetically improved 'post-human' elites. Obviously, in such a society, ordinary humans wouldn't be regarded as the equals of those produced through genetic manipulation."

The religious Right has eagerly embraced Smith's view of a pervasive animal rights–bioethics–transhumanist conspiracy to enslave humanity 1.0. The TechnoSapiens conference used a version of Smith's "The Transhumanists" as its motivating document.

In the Midwest the base for Christian Right bioethics is Chicago's Center for Bioethics and Human Dignity (CBHD), led by John Kilner, chair of ethics at Trinity International University. In 2003, Kilner and his CBHD colleague C. Ben Mitchell published "Remaking Humans: The New Utopians Versus a Truly Human Future." In addition to the charge that transhumanists hate humanity and are dangerous totalitarians in disguise, Kilner and Mitchell make clear another, specifically Christian objection: "Much of what the Transhumanists long for is already available to Christians: eternal life and freedom from pain, suffering, and the burden of a frail body. As usual, however, the Transhumanists—like all of us in our failed attempts to save ourselves—trust in their own power rather than God's provision for a truly human future with him." Human enhancement is seen as a distraction from the Christian promises of salvation in the afterlife.

In Washington, D.C., the locus of religious conservative bioethics is the Ethics and Public Policy Center (EPPC), dedicated to reinforcing "the bond between the Judeo-Christian moral tradition and the public debate over domestic and foreign policy issues." EPPC's "BAD" (Biotechnology and American Democracy) Project is headed by Eric Cohen, who works for Kass's President's Council on Bioethics as a senior research analyst. BAD's journal, *New Atlantis,* publishes attacks on artificial intelligence, nanotechnology, biotechnology, and reproductive technology and printed an anti-life-extension article from Kass in its premier issue.

JUST SAY NO TO A BRAVE NEW WORLD

With the extraordinary 2000 appointment of George Bush II by the U.S. Supreme Court, these institutions of the religious Right have gained unprecedented access and influence in federal health care policy, vetting appointees to science policy posts for their political and religious correctness and winning significant legislative and regulatory victories for "fetal rights," abstinence education and other issues of biblical concern. But they would not be half as influential if not for their eager secular conservative allies.

One such secular ally is Francis Fukuyama. When Fukuyama takes the train from Johns Hopkins University, where he teaches international relations, down to D.C., his think tanks are center-Right groups like the New America Foundation and the National Endowment for Democracy. And the bioLuddite worldview Fukuyama lays out in *Our Posthuman Future* is also secular. At least, on its face.

Fukuyama achieved intellectual notoriety in the early 1990s by arguing in *The End of History* that the collapse of communism marked the ascendance of democratic capitalism as the endpoint of human civilization. By the late 1990s, even Fukuyama had become convinced that his thesis was wrong, though not because he had discovered the joys of Swedish social democracy. Fukuyama had instead discovered human enhancement technologies and became convinced that they might inspire new, disastrous projects to improve human nature.

For Fukuyama and most social conservatives, utopian visions of radical improvement of the human condition, social or biological, lead inexorably to the gas chamber and the gulag. In the imagination of the anti-utopians every concentration camp guard was motivated by a vision of the New Man, while they ignore the utopian dimensions of the American civil religion. As Russell Jacoby notes, "the bloodbaths of the twentieth century can be as much attributed to anti-utopians—to bureaucrats, technicians, nationalists and religious sectarians with a narrow vision of the future." The conservative anti-utopians deny the possibility of liberal utopias, and of radically liberal projects like a transhuman democracy.

For Fukuyama any effort, pharmaceutical or genetic, to meddle with our ineffable humanness, our "Factor X," is by definition totalitarian. Fukuyama is admirably bold in refusing to provide any clue as to what

Factor X might be, however: "Factor X cannot be reduced to the possession of moral choice, or reason, or language, or sociability, or sentience, or emotions, or consciousness or any other quality that has been put forth as a ground for human dignity. It is all those qualities coming together in a human whole that make up Factor X." You see, if we were able to specify which traits were important to humanness, then we could make laws that preserved those and allow people to muck around with the rest. But since we don't know what Factor X is we can't change anything about human nature without accidentally breaking it.

Fukuyama next asserts that all people are currently created equal and treated equally before the law, and that enhancement would threaten this equality. Liberal bioLuddite Bill McKibben makes the same point in *Enough: Staying Human in an Engineered Age,* when he says, "The political equality enshrined in the Declaration of Independence can't withstand the destruction of the idea that humans are in fact equal." Here Fukuyama and the other human-racists have a point—social solidarity will be severely tested in the coming century by an increasingly diverse transhuman society, and advocates of transhuman democracy will need to build a new social solidarity and transhuman citizenship based on personhood. But human rights and legal equality depend as little on the homogeneity of species-typical abilities as they do on the homogeneity of skin color or height. As libertarian transhumanist writer Ron Bailey says:

> if we somehow reverted to the notion that human rights are based on physical equality, political equality would not long stand. Political equality has never rested on claims about human biology. After all, humanity had the same biology we have today during the long millennia in which slavery, patriarchy, and aristocratic rule were social norms. With respect to political equality, genetic differences, even engineered ones, are differences that make no difference.

Fukuyama is wrong about the inevitability of totalitarianism and a genetic caste system if we permit human enhancement, but he is right about two points in *Our Posthuman Future*. First, he is absolutely correct that the enhancement technologies *can* be regulated. Libertarian techno-utopians like Bailey imply that technology is somehow autonomous of humanity and that we can only "go with the flow." Bans on technologies are supposedly pointless since they will probably still be available somewhere. But if people have to fly to the South Pacific and pay tens of thousands of dollars for treatments that could have been available in their hometown for a fraction of the cost, that will significantly hamper the proliferation of that technology. They will only be available to the wealthy.

Second, Fukuyama is correct in arguing against the libertarians that enhancement technology *should* be regulated. Unfortunately he has far more reasons than simple safety and efficacy for regulatory agencies to regulate and forbid technologies. Fukuyama, like all the bioLuddites, treats every hypothetical negative consequence from the use of technology with great gravity, while dismissing as hype all the possible benefits.

BILL MCKIBBEN, THE UNABOMBER AND RENOUNCING PROGRESS

Kass and Fukuyama don't have any beef with industrial progress in general. They aren't tree-huggers. But the religious and political conservatives are finding common cause with the more traditional Luddites like Bill McKibben, author of the anti-transhumanist book *Enough*.

The basic point of *Enough* is that McKibben is satisfied with four-score years of life, with the current technologies of modern medicine, with the capacities of his brain, and with the world's level of economic development, and he thinks the rest of us should be also. He calls for the world to emulate the example of the Amish or Tokugawa Japan, and turn our back on further progress in order to contemplate and appreciate the virtues of the things we have. We all need to accept, he says, "that as a species we are good enough. Not perfect, but not in need of drastic redesign. We need to accept certain imperfections in ourselves in return for certain satisfactions. . . . We don't need to go post-human, to fast-forward our evolution, to change ourselves in the thoroughgoing ways that the apostles of these new technologies demand."

McKibben is most enraged by the transhumanist claim that we will actually find deeper, vaster, more awesome pleasures in enhanced life. How could a mind possibly experience anything more profound, have a sense experience more rich or think a thought more complex than those we have access to today? Living longer, healthier, smarter and happier will sap all meaning from our existence. Sunsets will be ho-hum when we can see a trillion more.

In *Enough* McKibben obliquely charges that the transhumanists might use terrorism against Luddites, then acknowledges we are generally "starry-eyed but peaceful." The comment is especially odd since it is precisely the radical wing of bioLuddites who have committed the only acts of terrorism in this struggle.

Take, for instance, the anarcho-Luddite Ted Kaczynski, the Unabomber. Between 1978 and 1996 Kaczynski mailed 16 bombs to targets in academia, killing 3 and maiming 23. He used his bombings to blackmail the media into publishing his 35,000-word manifesto in which he specifically addresses the need to dismantle medicine along with all other parts of industrial civilization, because of the threat from human genetic manipulation: "Man in the future will no longer be a creation of nature, or of chance, or of God (depending on your religious or philosophical opinions), but a manufactured product. . . . The only code of ethics that would truly protect freedom would be one that prohibited ANY genetic engineering of human beings." For Kaczynski the principal argument for destroying technological civilization was to stop genetic enhancement: "You can't get rid of the 'bad' parts of technology and retain only the "good" parts. . . . Clearly you can't have much progress in medicine without the whole technological system and everything that goes with it. . . . The temptation presented by the immense power of biotechnology would be irresistible, especially since to the majority of people many of its applications will seem obviously and unequivocally good (eliminating physical and mental diseases, giving people the abilities they need to get along in today's world)."

When Kaczynski was tried he rejected the insanity defense, and indeed his violent actions were consistent with his Luddite beliefs. For the human-racists in the right-to-life movement, killing an abortion doctor is the same as assassinating death camp doctors at Auschwitz—a moral obligation. For a serious anarcho-Luddite like Kaczynski, who believed that enhancement technologies will lead to the enslavement of humanity, the only moral course is to stop technological progress by all means necessary.

BILL JOY, RELINQUISHMENT AND THE PRECAUTIONARY PRINCIPLE

In Greg Bear's 1984 novel *Blood Music,* a scientist accidentally unleashes intelligent, self-replicating nanomachines that proceed to eat the planet. It actually turns out to be kind of OK in the novel since everybody is still virtually alive in the ooze. This apocalyptic scenario has become known as "gray goo." Once the bioLuddites heard about it from science fiction and transhumanists, gray goo became one of their many nightmares, along with genetic and nanomaterial pollution, runaway artificial intelligence, bioweapons and apocalyptic struggles between humans and posthumans.

The bioLuddite discovery of the threat from emerging technologies can largely be traced to an essay by Bill Joy. Until recently, Joy was the chief technologist and co-founder of Sun Microsystems and inventor of the computer language Java. In April 2000, he published a Luddite jeremiad in the unlikeliest of places, the militantly pro-technology *Wired* magazine. In the piece Joy confesses to having developed a serious case of anticipatory doom in contemplating genetic engineering, nanotechnology and artificially intelligent robots. These three technologies have the common feature that they can potentially self-replicate. Guns don't breed other guns and go on killing sprees but gene-tailored plagues and nanophages can theoretically do just that.

Hearkening back to the peace movement's effort to ban all nuclear weapons, Joy's answer to these apocalyptic threats is "relinquishment" of genetic, molecular and AI research: "These technologies are too powerful to be shielded against in the time frame of interest. . . . The only realistic alternative I see is relinquishment: to limit development of the technologies that are too dangerous, by limiting our pursuit of certain kinds of knowledge."

Joy implies that such relinquishment could be accomplished by a grand moral attunement of the scientific community. But the environmental movement immediately seized on Joy's call for relinquishment as a call to apply their "precautionary principle." For instance, the radical environmental group Rural Advancement Foundation International promptly changed its name to the Action Group on Erosion Technology and Concentration (ETC), with the new mandate of fighting nanotechnology and genetic engineering.

The "precautionary principle" as used by ETC and the environmental movement is deceptively commonsensical—no technology should be used until its risks are fully assessed. But the devil is in the details of enforcement. When exactly are all the risks of any technology fully assessed, and what do you do in the meantime? The Turning Point ad campaign organized by a network of bioLuddite groups in 1999 noted that many technologies, such as cars and antibiotics, have had unintended consequences. So is the implication that we should have forbidden the use of cars and penicillin until we figured out that they might result in smog and antibiotic-resistant bugs? Risk assessment always involves balancing immediate testable risks and benefits against future hypothetical risks and benefits. Aspirin might never have passed FDA approval, but it turned out to have myriad benefits in reducing the risk of cancer, heart disease and stroke. In the long run won't antibiotics have saved more lives than they threatened?

In the case of molecular and transhuman technologies there are enormous risks and enormous benefits, and the bioLuddites overstate the risks while ignoring the benefits. They focus on the entirely hypothetical health risks from genetically modified foods, and then dismiss the idea that genetically modified food may be necessary to feed a world of 10 billion. Yet even if they concede these kinds of benefits, they can still argue for a ban on research because they believe no possible benefits can outweigh even the smallest chance of apocalypse. What Joy and ETC leave out of their equation is the precariousness of human existence in the universe *without* technological progress. SARS and AIDS show that naturally evolving pathogens appear to need no genetic help from terrorists to threaten humanity, and molecular medicine is our greatest weapon in that fight. Our otherwise silent galaxy suggests that asteroid impacts, solar radiation and numerous other cosmic threats are waiting to wipe out this small accidental growth of intelligence. In his 2003 book *Our Final Hour*, British astronomer Martin Rees gives mankind a 50–50 chance of surviving the coming natural and man-made disasters of the twenty-first century. Without technologies that allow us to defend ourselves against these natural and man-made risks, to diversify into more durable and resilient bodies and spread beyond this vulnerable planet, human-spawned intelligence will have a much shorter run.

Technology bans would also leave us helpless in the face of those who developed proscribed technologies for military or terrorist purposes. We will need increasingly sophisticated detectors of and defenses against genetic, molecular and AI-based weapons. We will need to develop a nano-immune system to detect and counteract accidental and malicious nanotechnology, just as we have developed global networks to protect against computer viruses.

The risks of human enhancement are also fundamentally different from those of gray goo, supercorn and Terminator robots. It is unlikely that a genetically engineered human, or even two, will ever escape into the woods and eat the planet. Genetically or cybernetically modified humans are no more likely to be eco-destructive than humanity versions 0.1 through 1.0. On the contrary, transhumans with longer lives and greater intelligence would be more likely to foresee and avoid ecological destruction, and to have the technological capabilities to repair the damage already done. New genetically engineered crops and nanotechnology will make possible more ecofriendly agriculture and manufacturing. People who expect to be around in 200 years won't want to foul their own nests.

DEEP ECOLOGY AND THE DICTATES OF MOTHER NATURE

Left-wing bioLuddites are generally not traditionally religious. They don't base their opposition to technology on biblical edicts or God's plan for man. But thanks to New Age counterculture and radical environmentalism, political progressives have been infused with pre-humanist beliefs about the sacredness of nature, a belief system known as "deep ecology." Deep ecology was first articulated by the philosophers Arne Naess and George Sessions in the 1970s. But it began to spread with the growth of radical environmentalist groups like Earth First! The core of the deep ecology platform is the assertion that "The well-being and flourishing of human and nonhuman life on Earth have value in themselves. These values are independent of the usefulness of the nonhuman world for human purposes." Consequently, "Humans have no right to reduce this richness and diversity except to satisfy vital needs." Deep ecology has merged with pagan and New Age beliefs about nature spirits and the interdependence of all beings, but added a misanthropic undercurrent. In order to reduce humanity's excessive interference with the nonhuman world there must be "a substantial decrease of the human population."

In the 1980s, like a lot of progressive Buddhists, I was drawn to the alleged pagan–Buddhist syncretism of deep ecology and I was excited by the emergence of the European Green parties as the vehicle of this new spiritual politics. Slowly I came to realize that deep ecology was profoundly un-Buddhist and anti-humanist. As Buddhist philosopher John McClellan argued in his seminal essay "Nondual Ecology," a truly Buddhist outlook doesn't partition the world into natural (ecosystems) and unnatural (technology), but embraces both nature and technology as expressions of the evolutionary process. The technobiotic world created by human beings, argued McClellan, should be as embraced and celebrated as the trees and animals with which it co-evolves.

Since only human minds make distinctions like valuable and worthless, only human beings can give meaning and value to nature and technology. If all humans disappeared from the planet the ecosystem would continue, but the world would be more or less meaningless, at least until dolphins and apes evolved more abstract thought. For the deep ecologists, however, an Earth without humans would be as good or better.

Like the Christian Right, deep ecology calls for a return to natural law. Instead of seeing the development of all minds, all persons, as the Good, we are supposed to follow the dictates of Nature. Instead of human beings creating meaning in an otherwise meaningless universe, the universe already has a plan and it doesn't care about us. If Nature's plan requires that the human population be reduced to only half a million, as advocated by some Earth Firsters, so be it. Industrial civilization must be uprooted, and humans must return to hunting and gathering.

Imbued with these anti-human, deep ecological views, animal liberation and anti-GM food groups have no qualms about acts of violence against labs making new medicines or crops. Before 9/11 the majority of terrorist acts in the United States were being committed by deep eco-activists. In 2002 the FBI estimated that the Animal Liberation Front and Earth Liberation Front had together committed more than 600 criminal acts in the United States since 1996, resulting in damages in excess of $43 million. Noting that European fascism also had a nature-worshipping, anti-modernist wing, philosopher Greg Pence has dubbed deep ecology "ecofascism."

The moderate environmentalist and social justice movements have been fighting a rearguard defense against the spread of deep ecology for fifteen years. The U.S. Greens were polarized from the beginning by the condemnation of deep ecology by anarchists and socialists like Murray Bookchin. Reproductive rights activist Betsy Hartman and other progressives have struggled against the spreading influence of deep ecology's

Malthusian misanthropy, which gives warrant to coercive population control programs restricting the reproductive rights of developing world women.

Today Earth First! is led by a more politically correct leadership, less inclined to praise AIDS as Mother Nature's revenge on the human cancer, which has sought alliances with trade unions and anti-racist groups. But the influence of deep ecology is still pervasive throughout the liberal-Left and is found now in the writing of some of the most prominent leaders of the anti-human enhancement groups. One such deep ecologist is Andrew Kimbrell, the former policy director for Jeremy Rifkin, who went off to found the Washington, D.C.–based International Center for Technology Assessment. Most of Kimbrell's energies have been devoted to attacking genetically engineered crops, but he has taken time out to write *The Human Body Shop,* an attack on the alleged commodification of organs and tissues that he sees as "desacralizing" the human body.

Mainstream environmental groups are also beginning to line up with the opponents of transhuman technologies. Carl Pope, the director of the Sierra Club, used his address to the 2001 meeting of the National Abortion and Reproductive Rights Action League (now called NARAL Pro-Choice America) to urge the gathered pro-choice activists to support restrictions on parents' rights to germinal choice. The ecological think tank Worldwatch Institute devoted a 2002 issue of its magazine to a dozen articles opposing cloning and human genetic engineering, written by McKibben, Fukuyama and prominent feminist and environmental writers. Testifying before the U.S. Congress in 2002 in support of a ban on the use of cloning in medical research, Brent Blackwelder, president of Friends of the Earth, said, "The push to redesign human beings, animals and plants to meet the commercial goals of a limited number of individuals is fundamentally at odds with the principle of respect for nature."

One corollary of this deep ecological faith is that human beings can never be wise enough to improve on Nature, just as the theists assert that humans cannot and should not try to improve on God's Creation. Genomes are too complicated and mysterious to engineer, and attempts will lead to unpleasant, unintended consequences (think Tower of Babel and the untimely melting of Icarus's wings).

Another corollary of the deep ecological disdain for human science is suspicion toward scientific medicine. Granted, much medical practice is only barely more scientific than homeopathy or faith healing. But there is a difference between believing that modern medicine is not fully validated and should be used conservatively, and believing that all modern medicine makes people sick and should be avoided. This latter view, articulated by writers like Ivan Illich, carries over to a suspicion of the promise of human enhancement. Just as no number of studies demonstrating the safety of vaccinations can penetrate a natural health advocate's anti-scientism and convince them that vaccines are safe, no stack of clinical trials can demonstrate to bioLuddites that tinkering with the human genome could ever be safe. Arguing about the safety of technology with those under the sway of deep ecology is as pointless as arguing about the existence of God with the believer.

We certainly need to take every reasonable precaution as we set about making the universe a friendlier place for intelligence. I also believe we all should have reverence and awe for the universe. But humanity, its works and its tremendous promise are among the most awesome things of this universe. There are no sacred or "natural" limits to our taking control of our biology or ecology. There is no "natural" way to have a baby, grow corn, manage a forest or die. There are only ways we choose, some of which work better than others.

Left-Wing BioLuddites

Most political radicals embraced science and technology in the eighteenth and nineteenth century. But another group of radical critics have seen technology as an intrinsic part of the evils of capitalism, patriarchy and racism. Since World War Two techno-skepticism has come to replace techno-optimism on the Left. Now the Left techno-skeptics are turning their attention to biotechnology and nanotechnology, which they see as contributing to corporate and military power, racism, inequality and discrimination against women and the disabled.

THE RISE OF LEFT BIOLUDDISM

Political scientist Langdon Winner earns high marks with me for being the first person to warn the U.S. Congress of the threat from posthumanity. Winner testified before the House Science Committee on April 9, 2003, along with computer scientist Ray Kurzweil and nanotechnologist Chris Peterson. The hearing addressed funding for nanotechnology research. In response to a question about when there would be greater-than-human intelligence, Winner sternly intoned, "I hope never. One of the concerns about nanotechnology and science and engineering on this scale is that it is plowing onward to create a successor species to the human being. I think when word gets out about this to the general public they will be profoundly distressed. And why should public money be spent to create an eventual race of posthumans?" To which Ray

Kurzweil responded that "I would define the human species as that species that inherently seeks to extend our own horizons. We didn't stay on the ground, we didn't stay on the planet, we're not staying with the limitations of our biology."

Winner is probably the most famous, and one of the most sophisticated, of the Left Luddite theorists. In his classic *The Whale and the Reactor: A Search for Limits in an Age of High Technology,* he makes a careful argument that "artefacts have politics," that the power relations of society are designed into technologies. According to Winner, modern technology, selected for and designed under the thumb of corporations and the military, encourages centralization, hierarchy and the concentration of power. Some tools, such as nuclear weapons, have few positive uses, and necessitate authoritarianism. You don't want to leave nuclear weapons without a heavy guard, for instance, in case they are stolen. Other technologies, such as the crossbow, are very open to alternative uses, and can be turned against the powers that be. The crossbow was adopted by rebellious and larcenous European peasants because it allowed common people on foot to pierce the armor of knights on horseback before they came within cudgeling range. According to Winner, to create a more democratic society we need to democratize technology, from the shop floor to federal science bureaucracies.

Winner does not reject all technological progress. He calls for social reforms and careful consideration of social consequences as technologies are deployed, not knee-jerk bans. Unfortunately the subtlety of Winner's argument is often lost on those who use him in left-wing attacks on technology. For run-of-the-mill Left Luddites all technologies are bad because they are produced by corporations and unequal societies, and all technologies will therefore serve the interests of corporations, men and the rich. Therefore all technologies must be opposed.

Why did this end up being the default viewpoint of political progressives? The hostility of religious conservatives and ecomystics to technologies of human enhancement is understandable in light of the long struggle of science, reason and humanism against religious orthodoxy.

The political message of liberty and equality was closely tied to the advocacy of science, reason and free enquiry, and both faced suppression by the church and aristocracy. The revolutionary changes wrought by technology were believed capable of making possible a new, better and more prosperous society. The idea of political progress became tied to the idea of technological progress, and nineteenth-century socialists, feminists and democrats generally championed reason and science. Eighteenth-century radicals like Joseph Priestley and the Marquis de Condorcet pursued scientific investigation while championing democracy and freedom from religious tyranny.

The ideas of natural selection and evolution were used by the Right to argue that the rich were simply the most fit, but also by the Left to validate the idea of social progress. The Oneida community, America's longest-lived nineteenth-century "communist" group, practiced extensive eugenic engineering through arranged breeding. For American socialist Edward Bellamy, author of the popular utopian vision *Looking Backward,* and as for the Fabian Socialists in the United Kingdom, socialism would be the result of a painless evolution of industrial society.

Karl Marx and Friedrich Engels believed more pain and conflict would be involved, but agreed about the inevitable end. Marxists argued that the advance of technology laid the groundwork not only for the creation of a new society, with different property relations, but also for the emergence of new human beings reconnected to nature and themselves. At the top of the agenda for empowered proletarians was "to increase the total productive forces as rapidly as possible." The exiled Russian revolutionary Leon Trotsky expressed the transhumanist aspirations of the Marxist Left when he wrote that, after the Revolution:

> Man at last will begin to harmonize himself in earnest. . . . He will try to master . . . breathing, the circulation of the blood, digestion, reproduction, and . . . subordinate them to the control of reason and will. . . . The human species . . . will once more enter into a state of radical transformation, and, in his own hands, will become an object

of the most complicated methods of artificial selection and psycho-
physical training. This is entirely in accord with evolution.

Meanwhile the American Progressive movement and New Deal
built hydroelectric dams and federal science agencies, and set up regulatory labs to employ scientists to serve the public interest. The 1945
report to President Roosevelt that laid the groundwork for the estab-
lishment of the National Science Foundation, *Science: The Endless Fron-
tier,* attributed American prosperity and longevity to "the free play of
initiative of a vigorous people under democracy, the heritage of great
national wealth, and the advance of science and its application. Science, by itself, provides no panacea for individual, social, and economic ills. . . . But without scientific progress no amount of achievement in other directions can insure our health, prosperity, and security
as a nation in the modern world."

So why did the partisans of the idea of political progress, of a more
just society, become estranged in the late twentieth century from the
idea of technological progress? Why are so many contemporary social
democrats, feminists and Greens suspicious and hostile to biotechnologies, computers and science in general? The answer probably starts with
the romantic traditions that grew up in reaction to modern capitalism,
the shadow twin of the techno-optimist progressives. Pastoralist visions
of a deindustrialized socialism, pseudo-science, spiritualism and back-to-
land communalism were all tied up for bohemian radicals with opposition to capitalism. In fact, in the *Communist Manifesto* Marx and Engels
specifically warn against clerical, aristocratic and petit-bourgeois socialists who advance pastoralism and pre-industrial society as an alternative
to capitalist progress.

But it wasn't until World War Two that the rationalist, industrializing Left lost out to the romantic bohemians. The Left's interest in
reengineering the nature of Man was silenced by Nazi eugenics. The
gas chambers revealed that modern technology could be used by a
modern state for horrific uses, and the atom bomb posed a permanent

technological threat to humanity's very existence. The ecological movement suggested that industrial activity was threatening all life on the planet, while the anti-nuclear-power movement inspired calls for technology bans. The counterculture attacked positivism, and lauded pre-industrial ways of life. Deconstructionists and post-modernists cast doubt on the "master narratives" of political and scientific progress, while cultural relativists attacked the idea that industrialized secular liberal democracies were in fact superior to pre-industrial and authoritarian societies.

While the Progressives and New Dealers had built the welfare state to be a tool of reason and social justice, the New Left joined cultural conservatives and free-market libertarians in attacking it as a stultifying tool of oppression, contributing to the general decline in faith in democratic governments. In *One Dimensional Man,* Herbert Marcuse, theoretician for the New Left, wrote that "technological rationality" was "the great vehicle of better domination, creating a truly totalitarian universe." Todd Gitlin, the former president of the Students for a Democratic Society, commented that while "an orientation toward the future has been the hallmark of every revolutionary—and for that matter liberal—movement of the last century and a half," the New Left's pastoral brand of anti-futurist utopianism suffered from "a disbelief in the future. . . . We find ourselves incapable of formulating the future."

As the Left gave up on a sexy, high-tech vision of a radically democratic future, libertarian champions of corporate capitalism became associated with technological progress. Techno-enthusiasm on the Left was supplanted by pervasive Luddite suspicion about the products of the corporate consumerist machine. Celebrating technology was something GE and IBM did in TV ads to cover up their complicity in napalming babies. Activists fought the machine. So the first reaction of the modern Left to all the new biotechnologies, especially any technology having to do with genetics, was suspicion and loathing. That has carried over to a near universal disdain and hostility toward the prospects for human enhancement among political progressives.

PROTECTING THE POOR FROM TECHNOLOGY

To the Left Luddites, human enhancement is a bourgeois waste of resources in a world of crushing poverty and need. Socialist writer and building manager George Scialabba, reviewing Fukuyama's *Our Posthuman Future* in the liberal journal *American Prospect,* writes: "The best reason of all not to press forward into the posthuman future . . . goes unmentioned in this book. It's that the enormous resources required could be put to much better use helping the many people who do not now enjoy a human present." He notes that 1.5 billion people don't have access to clean water, 1.25 billion lack adequate housing and a half billion consume too few calories. One billion have no health care, and many millions of children are malnourished. "The UN estimates that all these needs could be met, at a basic level, for a yearly expenditure equal to 10 percent of the recently proposed U.S. military budget—or slightly less than Americans and Europeans spend annually on pet food and ice cream."

Of course Mr. Scialabba is right, but why start by banning Americans' use of human enhancement medicine, which might eventually benefit the poor, instead of American pet food and ice cream, which will never benefit the world's poor? If we agree that some sacrifices need to be made to correct these monstrous wrongs, surely our rights to pets or ice cream are more expendable than our rights to control our own bodies and reproduction. It is clearly insane to live in a world where everyone could have food, shelter and clean water, but don't. But the knee-jerk Left has neither grappled with the right to bodily autonomy nor shown how banning life extension medicine or germinal choice or cybernetic implants would lead to a redistribution of those dollars to the world's poor.

More thoughtful Left critics argue that transhuman technologies need to be banned because they will make society more unequal. Transhuman technology will lead to a genetic caste system, says the Center for Genetics and Society (CGS) in Oakland, California, the spearhead of the leftist caucus of bioLuddism. The CGS started as an ad hoc lobby in the late 1990s raising the alarm about "technoeugenics," that is, cloning, designer babies and inheritable genetic modifications, which they believe will lead to a permanent division between the "GenRich" and "Gen-Poor," terms borrowed from Lee Silvers' pro-germinal choice book *Remaking Eden*. Then in 2001 CGS was adopted by the Tides Foundation, a large progressive meta-philanthropy in San Francisco, allowing it to hire staff and begin to organize in earnest. CGS has helped organize the campaign for an international treaty to ban cloning and inheritable genetic modification. CGS staff now routinely write op-eds for the media attacking transhumanists and advocates of germinal choice, and accusing us of advocating "eugenics."

EUGENICS AND ACCESS TO GERMINAL CHOICE

Whether germinal choice really is eugenics depends on one's definition of eugenics. The eugenics movement that spread across Europe and the United States before 1945 encouraged selective breeding and was responsible for the mandatory sterilization of criminals, the poor, the disabled and dark-skinned people based on unscientific theories. There are very few advocates of this older eugenics around today, and to the extent that anyone advocates racist, classist or authoritarian ideas they are to be despised.

But if eugenics includes believing that individuals, free of state coercion, should have the right to change their own genes and then have children, then the advocates of human enhancement and germinal choice are indeed eugenicists. If eugenics also includes the belief that parents and society have an obligation to give our children and the next generation the healthiest bodies and brains possible, then most people are eugenicists. Once safe, beneficial gene therapies are available parents will feel the same sense of obligation to provide them for their kids as they do a good education and good health care. As bioethicist Arthur Caplan has said, "many parents will leap at the chance to make their children smarter, fitter and prettier. . . . They'll slowly get used to the idea that a genetic edge is not greatly different from an environmental edge."

On the other hand, if eugenics is authoritarian "genetic correctness," it is precisely the bioLuddites who are today's eugenicists. The bioLuddites are the ones who want laws on what kind of children we can and can't have, who want to forbid people from controlling their own bodies and reproductive choices. The big difference is that while the American eugenicists were using profoundly anti-democratic classist and racist theories to try to *improve* life for the next generation, the bioLuddites are motivated by their anti-democratic human-racism to want to *stop* people from improving the lives of the next generation.

Nonetheless, the CGS is right that widespread individual use of germinal choice could make society more unequal. But so can the use of every new technology from cars to computers. The answer isn't to stop technology but to change the system that unequally distributes technology. Rich kids have access to better computers, cell phones, vitamins, health care and jobs than do poor kids. The answer is not to take away these advantages, but to make sure that everybody has better access to all of these riches. Technology doesn't make people unequal; power and wealth do.

The ETC Group is another international Left Luddite nonprofit based in Canada that has focused on nanotechnology and biotechnology. In a 2003 report attacking the National Science Foundation's initiative to "improve human performance" through the convergence of nanotechnology, biotechnology, information and cognitive science (NBIC), ETC expressed the Left Luddite position clearly: "Powerful new technologies in the hands of those who have benefited by—and perpetuated—inequality, will exacerbate the gap between rich and poor. It is irresponsible to contemplate such a massive technological change without first enacting positive social change." In other words, rather than fight for an egalitarian society that better distributes the benefits of technology, all technological progress should stop until we have a democratic and egalitarian society. Liberal theorist Ronald Dworkin responds to their argument in his book *Sovereign Virtue* that "We should not . . . seek to improve equality by leveling down. . . . The remedy for injustice is redistribution, not denial of benefits to some with no corresponding benefits to others."

While the Luddites overestimate the power of technology to create inequality, they underestimate the inevitability of popular demands for more equal access to technology. The weakness of egalitarian politics in the United States makes the U.S. Left especially prone to Luddite defeatism. While every other industrialized country has a powerful social democratic party and national health care systems politics in the United States is a choice between liberal and reactionary corporate control while 40 million Americans are still uninsured.

The good news is that demands for universal access to the benefits of human enhancement and life extension are about to make universal health care unavoidable in the United States. Reflecting on Lee Silvers' naïve, and now frequently quoted, prediction that germinal choice will lead to a division between the genetic haves and have-nots, conservative Adam Wolfson retorted in the *Public Interest* that "if genetic enhancement were to become possible, democratic publics would take to the streets with knives and guns before allowing Silvers' scenario to come to pass. The lower and middle classes would insist that their children be provided with the same eugenic enhancements available to the children of the rich. In time, the U.S. government would subsidize eugenic programs, not to create an overclass but to preserve equality, to elevate everyone's natural endowments." The point of this book is to argue for precisely such a movement, a movement that progressives will be forced to embrace once they have exhausted the dead-end of bioLuddism.

DESIGNER BABIES AND THE VICE OF GENETIC GAMBLING

Imagine a parent who made every decision for their children by rolling a pair of dice. How many years should you attend school? Come on double sixes! Oh, sorry, snake eyes—just two years for you. Such a parent would be considered unfit for parenting. Yet, when it comes to the genetic crapshoot, parents who want to assure the best for their child, who want them to be as healthy and able as possible, are the ones on the defensive, having to prove that they really love their children and don't think of them as a commodity. Bioethicist Thomas Murray, in *The Worth of the Child,* writes, "Good families are characterized more by acceptance than control." But good families are not characterized by indifference to whether a child is healthy or not, or bright or not.

Parents deserve to exercise all the genetic choices science can provide, and they deserve the benefit of the doubt that they will make choices that enrich their children's lives. Who more than the parent, who will have to raise the child, would care more about the many trade-offs, risks and unknowns that will go into each decision?

While dressed up with Marxian rhetoric, the Left bioLuddite opposition to commodified designer babies is just another doctrine of religious natural order. Every little baby is a precious gift, and to want to improve them in any way is an insult to Yahweh or Gaia or the uncommodified parent–child relationship. But that logic equally applies to birth control, as the Vatican consistently points out, as well as to prenatal care of any kind. If a mother takes prenatal vitamins to prevent spina bifida does it means she would love a baby born with spina bifida less? If she goes to an expensive obstetrician because she wants to ensure the health of the baby, has she commodified the baby? Will she shun it if it turns out to have been a failed investment? Obviously not.

A society with designer babies will supposedly be obsessed with genetics and forget about the importance of parenting. In fact, just like birth control, abortion and family planning, germinal choice is likely to increase the fit between kids and the desires of their parents, making both happier in the long run. Transhuman technologies will actually strengthen families by helping to ensure that every child is an even more wanted child, including all their various traits.

Critics fret that the new family structures that reproductive technology makes possible might harm kids. There is substantial evidence that having at least two caregivers in a house is better for children than single-parenting. But all the rest of the vague anxieties about "unnatural" families have no empirical support. More than a decade of studies of children raised by gay and lesbian couples find that they are just as psychologically well-adjusted as children raised by straight parents. A decade of large studies also find that children conceived with artificial insemination and other infertility treatments, "children of choice," are just as well-adjusted, and in some cases more loved, than "children of chance."

For instance, University of London researcher Susan Golombok has studied hundreds of families with children produced by invitro fertilization and donor insemination, up through age 12, comparing them to families that have adopted or conceived naturally, and finds no evidence that reproductive technology contributes to lower-quality parenting, high expectations or a "commodified" parenting style. She has found repeatedly that "the quality of parenting in families with a child conceived by assisted conception is superior to that in families with a naturally conceived child" and that the children of choice are as or more psychologically healthy. Parents who had used artificial methods expressed more warmth toward their children than the parents in the other two groups, and their children expressed more warmth for their fathers and were less likely to engage in physical aggression toward their peers. Some studies have found that children produced through artificial means have superior cognitive and intellectual abilities, and none have found that they have any deficits.

Critics go to quite absurd lengths to portray parents who want to use reproductive technology as dangerous. Fukuyama suggests that fathers will feel incestuous toward daughters cloned from their wives. If it were in fact true that fathers are more likely to feel incestuous toward girls that resemble their wives then we probably should encourage everyone to switch kids at birth, since many uncloned daughters resemble their mothers. Or perhaps we should just set aside the vague, baseless anxieties about "unnatural" families as reasons for interfering with procreative liberty.

FEMINISTS AGAINST REPRODUCTIVE CHOICE

When my wife was pregnant with our first child we attended a lecture at the University of Chicago hospital where she planned to deliver. The lecture was on the history of electronic fetal monitoring, and the conclusions were startling. Study after study had shown that the electronic monitors could not be accurately read, provided no benefit for mother or baby, and increased the rate of unnecessary cesarean sections. But when we confronted the midwives at the hospital with this information, their hands were tied by administration's risk management policies. So we delivered both of our kids at home, and became well educated about the unscientific and harmful practices masquerading as obstetric medicine. My wife in particular became profoundly distrustful of modern medicine, and our disagreements were passionate.

But at 42 my wife wanted to have a third child, and found she couldn't conceive. During a round of fertility drugs and daily trips to the clinic she reread an old feminist tract about how all women who use reproductive technology are dupes of capitalist patriarchy. She turned to me at dinner that night and said, "I finally get what you and the transhumanists are about. This is really insulting. They think I'm not smart enough to use technology without being a brainwashed puppet."

It seems obvious to many that the right to use reproductive and germinal choice technologies would be included in our right to bodily autonomy, and specifically in our "procreative liberty," a phrase promoted by lawyer-ethicist John Robertson in *Children of Choice*. But this isn't clear to bioLuddite feminists. In 1993, Janice Raymond, an activist with the group FINRRAGE (Feminist International Network of Resistance to Reproductive and Genetic Engineering), wrote in her book *Women as Wombs:* "Technological reproduction completes the medicalization of sex begun in the nineteenth century. The sexual objectification and violation of women is made invisible because technological reproduction has turned medicalized pornography into education, made medicalized access to the female body acceptable, and transformed medicalized abuse into treatment. Technological reproduction is first and foremost about the appropriation of the female body."

Feminists have good reasons to be suspicious of the patriarchal practices of obstetrics, and the assumptions behind reproductive medicine. The feminist critique has led to calls for women to take more control over their reproductive decisions and be more assertive with the health care system. But suspicion and hostility toward medicine and science have led to a concerted feminist–Luddite attack on reproductive technology in books such as *The Mother Machine* by Gena Corea, *The Politics of Reproduction* by Mary O'Brien and the collections *Man-Made Women: How the New Reproductive Technologies Affect Women* and *Made to Order: The Myth of Reproductive and Genetic Progress.* In the feminist–Luddite critique, women who want to use reproductive technologies are victims of false consciousness, and giving them choices is actually bad for them. As Janice Raymond writes in *Women as Wombs,* "Feminists must go beyond choice and consent as a standard for women's freedom. Before consent, there must be self-determination so that consent does not simply amount to acquiescing to the available options."

Some feminists are now joining forces with the religious and environmental bioLuddites to oppose reproductive technology, cloning and germinal choice. Feminist authors Naomi Klein and Judith Levine, women's health activist Judy Norsigian and other prominent feminist activists joined the Rifkin-organized progressive bloc in opposition to the use of embryos in medical research, even though it meant joining forces with the right-to-life movement. Norsigian says that women can't ever give informed consent to genetic therapies since those risks can't be fully known.

In *The Other Machine,* a feminist *defense* of reproductive technology, Dion Farquhar calls this "fundamentalist feminism." She argues that the fundamentalist or bioLuddite feminists refuse to acknowledge any connection between a reproductive right to abortion, which they defend, and a right to use reproductive technologies, such as germinal choice, which they want to deny. If a woman has a right to kill her fetus or poison it with alcohol, that would certainly seem to cover a right to improve her own genome and then reproduce, or to clone herself, or perform genetic therapies on her unborn child for its benefit. If women should have the right to choose the father of their child, with his attendant characteristics, then they should be allowed the right to choose those characteristics from a catalog. The progressive bioLuddites, in their zeal to prevent the return of coercive eugenics, appear to have forgotten that the right to control our bodies, including our reproductive options, is so important that the state needs some very solid reasons before interfering with them. As Marcy Darnovsky of the Center for Genetics and Society says, "It will take focused effort to make it clear that altering the genes of one's children is not among the reproductive rights for which so many women and women's organizations have struggled."

LEGITIMATE REASONS TO RESTRICT REPRODUCTIVE RIGHTS

People should have a right to control their own genomes and have children without permission from the government. So I don't think we really can or should do much to prevent germinal choice that results from someone intentionally changing their reproductive cells with a legal gene therapy. But we can and should pass laws about what kinds of medical products, services and technologies can be sold. All genetic therapies on embryos, fetuses, children and adults, like all drugs and medical devices, should first pass through animal and then human trials, and be shown to be safe and effective before they are made available to the public. Human cloning and germline modifications should have to show in clinical trials that they do not cause genetic abnormalities in mammals. The first human trials of cloning or genetic modification could be restricted to parents who need to use the technique to have a child because a parent has a serious genetic disease. Then, after observing these first experimental clones and gene-tweaked kids, we could open the technique to all prospective parents who want to use it.

The length of time we need to wait to be certain of the safety of genetic modifications will quickly shrink as we build computer models of the human genome and the proteins and tissue engineering that it codes for. One of the larger "in silico biology" efforts is the international E. Coli Alliance, which has created a computer model of the single-celled organism and how its genes work together to create the organism. Now they can begin tweaking it in computer simulations.

With the exponential progress in gene scanning and computing power, a virtual model of genetic expression in the human body will not be far behind. Humans are only one order of magnitude more complex than *E. coli* (35,000 genes to *E. coli*'s 4403 genes), which is about five years in Moore's Law time. We are already testing virtual drugs against virtual tissues. The firm Entelos in California has built a virtual asthma patient using equations that model 7500 biological parameters relevant to asthma such as the effect of bronchial inflammation and the thickness of the mucous lining in the lungs. The Pharcyte Corporation has built a vast database of clinical trial data, so that pharmaceutical companies can use "computer-assisted trial design" to estimate best dosages and the likely outcome of trials. The firm Gene Network Sciences has built a virtual colon cancer cell, which simulates the expression of 5000 genes and proteins, includes all the scientific knowledge about how the cell works in a series of algorithms, and models every one of the cell's known drug targets. So we will soon be able to speed up approval of gene tweaks by simulating their effects on virtual humans. Artificial chromosomes may also turn out to be a safer alternative to existing genetic engineering techniques.

Some opponents of genetic engineering, however, argue that even if germline genetic modification appeared safe for its first generation, it should still be forbidden since we can never be certain that some future generation won't regret the choices we made. Francis Fukuyama told Salon.com, "It could be that you will try to enhance the child's intelligence, and it will turn out that you'll increase the susceptibility to certain kinds of cancers—but this won't show up until the child is 60 years old." That kind of concern is not a real risk of a real harm, and not legitimate grounds for interfering in the very real reproductive freedoms and obligations of parents. Any appeal to the dangers of germline modification beyond adolescence, not to mention hypothetical anxieties about effects on the next generation, are absurd since future generations will be upgrading whatever genome they get from their parents. We won't be selecting genomes for all time, but starter genomes for a child's first twenty years or so.

Some bioLuddites appear to assume that there are many parents eager to mentally and physically cripple their children. There certainly may be a few. In *What Sort of People Should There Be?* Jonathan Glover asks what we should do if a religious minority engineers a sign of their faith on their children's foreheads, or engineers their brains to prevent apostasy. Certainly we have a duty to all kids to make sure that they not be robbed of any abilities and choices through genetic engineering. Likewise it is our responsibility to make sure children aren't robbed of their fullest abilities and choices by lack of access to prenatal care, screening, nutrition, education or their parents' *lack* of germinal choices.

So what might we do with parents who wanted to use germinal choice to rob their children of intelligence, health or abilities? We already face this issue with a couple sets of deaf parents who have tried to use IVF and prenatal screening to select embryos with congenital deafness. These parents argue that deaf culture is equally valid and that they would be better able to raise a deaf child. According to the principle of "procreative beneficence," selecting for deafness is clearly unethical, since it robs children of a critical ability. Physicians should refuse to accede to such a request, and public and private insurance should refuse to pay for it. When the deaf child reaches maturity they should be able to sue their parents for damages. As Peter Singer argues, selecting a deaf child is as much an abuse of the child as the intentional deafening of a hearing child, since, either way, that child "will never be able to hear Beethoven, or a babbling brook, or listen to lectures and debates delivered in spoken languages, except in translation. The child will also be at a disadvantage in countless other ways in getting through life." Even if technology can make them hear later, they will have passed the critical threshold of brain development and will be unable to process speech. A germline therapy that deafened a child, regardless of its other benefits, would never be approved for use by federal regulators, so why would we permit parents to deafen a child through their choice among embryos?

I could be comfortable with laws against genetic interventions that rob children of abilities. But we probably value procreative liberty so highly, and set such a high bar on the state's interference in reproduction, that we will permit parents to choose deafness if they pay for it themselves and can find a willing clinic and doctor to perform the procedure. We may still wish to draw the line at choices that deny children longevity or intelligence. Or, since we permit women to smoke and drink during pregnancy, we may even decide to allow parents complete procreative liberty and germinal choice, and only try to dissuade them from bad choices with public shaming and education. In such a laissez-faire society I am confident that 99.99% of parents will use germinal choice technologies to enhance their children and not harm them. The risk to the one-in-a-thousand child harmed by parental choices may be an acceptable trade-off for respecting procreative liberty.

Later I outline which germinal choices I believe should be subsidized, which left to the market and which strongly discouraged. In general, however, once a germinal choice technology has been tested and shown not to cause a high rate of birth defects or health problems, I believe parents should be permitted to use it. Figuring out what are harmful choices for children in the future will become very complicated. If people choose to modify themselves to live underwater, with gills and flippers, and then choose to have children to share their underwater society, would this be child abuse or enhancement? It takes away some abilities but adds others. Trusting parents to navigate the increasingly diverse choices will be hard, but nonetheless essential in a free society.

A second class of harms that warrant restrictions on reproductive freedom, social harms, are more difficult to predict, but also far less urgent. An imbalance of males over females is often suggested as a reason to forbid parents the right to choose their children's gender. Indian parents have widely used ultrasound and amniocentesis to determine the sex of pregnancies and abort female fetuses, contributing to a sex ratio of 9 females for every 10 males. The Indian government outlawed the practice in 1994 and has become increasingly punitive against clinics that provide prenatal screening information. Authorities have orders to seize any ultrasound machine used to tell a mother the sex of her fetus.

But why exactly is it the Indian government's, or any government's, business to ensure gender balance in the population? The Indian government isn't actually preventing sex discrimination since female fetuses aren't persons and can't be discriminated against. It is not discrimination against ugly people if a woman decides to abort a child sired by an ugly man. Is it the government's responsibility to see that every heterosexual can find someone to marry? If so, women should file a class action suit in favor of sex selection for more boys since being 51% of the population makes it harder for them to find a mate.

In fact, efforts to forbid sex selection threaten women's reproductive freedom. The Indian law forbidding mothers from finding out the gender of their children also radically restricts their right to information about their pregnancy. In India both a woman and her clinic are subject to criminal penalties if she receives an ultrasound or genetic test on her fetus and she isn't older than 35, or doesn't have a history of disabilities or miscarriages. Which means women are not only forbidden from finding out the gender of their fetus but also from assuring themselves that it is healthy.

Patriarchal biases against girls are certainly distasteful, just as we may find it distasteful for a woman to abort a fetus because the father was the "wrong" color. But it is clearly preferable that parents have *wanted* children rather than *unwanted* children. The consequence of a lack of germinal choice in the developing world has been more underfed girls and female infanticide, not gender equality. Most parents in the developed world who choose their child's sex, at invitro clinics or through abortion, do so to balance their families, not because they favor boys. And sex selection in Japan actually tilts to a three to one preference for girls.

In almost all cases, all the social risks and the psychological harms alleged for kids who are designed, sex selected, cloned or enhanced are simply baseless speculation, and do not amount to concrete reasons to interfere with rights to control our own bodies. As lawyer and bioethicist John Robertson notes: "In the debate over germline genetic engineering, it is essential to focus in on the interests of the parents who want to use this technique. If it can be shown that parents' desire to use this technique is part of their reproductive, their procreative, or their parental liberty, then the government, the state, would have to come up with much more tangible evidence of harm if it wishes to interfere with or stop their use. If fundamental liberties are involved, speculative, hypothetical harms will not do as a justification for law."

We still have a ways to go to convince the world of the importance of procreative liberty. Most accept the use of germinal choice for therapeutic purposes, but reject cloning, sex selection and enhancement. A survey of Americans conducted by the Genetics and Public Policy Center at Johns Hopkins University in 2002 found that majorities favored genetic testing or engineering of embryos to avoid disease (see Table 9.1). But only a quarter favored pre-implantation genetic diagnosis for non-therapeutic reasons like gender, and only 20% favored genetic enhancement or human cloning. Yet the 20% accepting enhancement and cloning is already surprisingly large. That adds up to about 60 million Americans, about the same percentage who are evangelical Christians. As recently as 1993, only 10% of the public thought human cloning would be acceptable.

While the public is still fuzzy about the importance of procreative liberty, they are very clear about our right to control and modify our own body and then have children. While majorities opposed genetic enhancement in this survey, a majority (59%) also favored parents being able "to change their *own* genes in order to prevent their children from having a genetic disease," the inheritable germline therapy that the bioLuddites seek to ban.

TABLE 9.1   Approval of Germinal Choice Technologies
From a survey of 1,211 adult Americans in October 2002 conducted by the Genetics and Public Policy Center at Johns Hopkins University.

Technology	Approve
Pre-implantation Genetic Diagnosis (PGD) to avoid serious genetic disease	74%
invitro fertilization	72%
PGD to ensure child is a good tissue/blood match	69%
Prenatal testing for disease	66%
PGD to avoid a tendency to diseases like cancer	60%
Genetic engineering to avoid disease	59%
Animal cloning	37%
PGD to choose child's sex	28%
PGD to ensure child has desirable characteristics	22%
Prenatal testing for desirable traits	20%
Genetic engineering to create desirable traits	20%
Human cloning	18%

DIVERSITY, DISCRIMINATION AND THE TECHNO-FIX

Despite "backlash" and temporary reversals, the line of social advance carries us toward a wider tolerance, a more easy acceptance of more and more diverse human types.

—ALVIN TOFFLER,
Future Shock

Opponents of germinal choice often assume that choosing to have a boy or an able-bodied child will weaken rights for women and the disabled. This is fundamentally muddled thinking, based on the theory of fetal personhood. Killing a child because she is a girl or mentally disabled is murder and a hate-crime, but deciding not to have them in the first place is not unethical. Otherwise deciding not to have more children would also be murder.

Nor does it make sense to tie the rights enjoyed by a group in society to their number in society. Is the argument that if half the world was blind there would be more things to read in Braille? If half the world was in wheelchairs there would be more wheelchair access? If there were more women in the population they would have more rights? On those grounds disability rights activists should be lobbying against workplace safety laws so there would be more people maimed on the job, and feminists should be in favor of sex selection for female fetuses.

Critics of germinal choice also insist that a society that is healthier and more able will discriminate more against the disabled. A position paper from the Council on Responsible Genetics against germline engineering says, for instance, that "The cultural impact of treating humans as biologically perfectible artifacts would be entirely negative. People who fall short of some technically achievable ideal would be seen as 'damaged goods,' while the standards for what is genetically desirable will be those of the society's economically and politically dominant groups. This will only increase prejudices and discrimination in a society where too many such prejudices already exist."

But that is not what we have seen in this century. Precisely as the rates of disability in society declined, there was an increase in efforts to mainstream the disabled into society, provide them assistive technology and ensure they reach their fullest potentials. There is no trade-off between disability rights and policies permitting parents to select for healthier children or encouraging a reduction in disabilities. We can and will have both. The argument that germinal choice is anti-disabled or patriarchal is yet another manifestation of the left-wing bioLuddite error: fighting individuals' free use of technology instead of power relations and prejudices.

The Left bioLuddites respond that, even if we could work on both simultaneously, a society that allows disabilities to be corrected or that selects their kids' sex will spend less effort on making society welcome for women and people with disabilities. Genetic technology will make it seem that all social problems have a techno-fix, delaying needed social reforms.

What if black parents decide to give in to racism by giving their children a skin-lightening fix, allowing them to pass as white? Just ask people of color how many of them would choose to have kids that look white in order to allow those kids to "pass." If there is a gay gene, some religious conservatives might set aside their objections to germinal choice and decide to select for straight kids. That would be a shame. But is forcing homophobic families to raise gay children the best way to fight homophobia?

Control over human nature is unlikely to lead to neglect of environmental improvement. Society might just ramp up kids' intelligence instead of providing them with better-funded schools. But that wouldn't work very well, since smarter kids would only make the inadequacies of the schools more glaring. We will fix obesity genes, but people will still have to eat right and exercise. Fixes for lung cancer and skin cancer are unlikely to dry up our concern about industrial pollution and the ozone layer.

Another common concern is that germinal choice will increase genetic testing and discrimination. The movie *Gattaca* depicts a future in which genetic engineering of humans is widespread, and is accompanied by widespread testing for genetic aptitudes for various jobs. The protagonist is trying desperately to hide his genetic predisposition to heart problems so that he can pursue a career as an astronaut. The film's dystopian depiction of genetic discrimination is now widely cited as one of the convincing proofs that germinal choice is a bad idea. Setting aside the fact that astronaut-training programs are entirely justified in attempting to screen out people with heart problems, will germinal choice create a society where people are sorted into rigid castes on the basis of their genetics? Will people be denied jobs or insurance on the basis of their genetics?

Yes, possibly, and it has absolutely nothing to do with the availability of genetic enhancement since people are already being discriminated against on the basis of their propensities to disease. American insurance companies routinely deny coverage to people with preexisting conditions, and will want to make use of every new piece of diagnostic information in order to extend their predictive accuracy and profit margins. In a country where half the employees are required to submit to routine drug testing, genetic testing for disease is not much farther behind. The answer, again, is not to ban genetic tests or genetic enhancement, but to ensure the privacy of genetic information, and strictly control when genetic information can be used to make decisions in education and employment.

We need to return to a really basic idea: It is our duty and right to provide children with the healthiest and most able bodies we can. Bioethicist Peter Singer makes this point when he proposes that "For any condition X, if it would be a form of child abuse for parents to inflict X on their child soon after birth, then it must, other things being equal, at least be permissible to take steps to prevent one's child having that condition." In other words, the parent who doesn't take reasonable, available steps to ensure that their future child has normal abilities is as morally culpable for their child's disabilities as the parent who causes those disabilities after birth.

Australian bioethicist Julian Savulescu sharpens the point, and applies it to germinal choice, with his principle of "procreative beneficence": We have an obligation to choose the child, among the possible children we could have, which will have the best life. That means we are obliged not only to choose children without disabilities, but also to create enhanced children, so long as the enhancements are safe and available. If we can choose between an embryo with a projected 150-year life span and one with only a 70-year life span, we are obliged to pick the former, just as we are obliged to pick a 70-year life span over a 10-year life span. We have this general obligation as a society, to work to make sure that the next generation is as healthy, intelligent and able-bodied as possible, just as we have it as individual parents. The best way that society can fulfill that obligation is to make sure that all parents are able to provide the good things in life for their children, from nutrition and education to net access and optimal genomes.

Whenever I meet a parent of a disabled child who is outraged that germinal choice is "discrimination against the disabled" I ask them if they would fix their kid's disability if they could. The answer is invariably yes. My daughter suffers from Tourette's syndrome and my son from ADD. Both conditions are linked to the same gene, a gene they probably got a double whammy of from my wife and me. If we had had the option to snip out the propensity to ADD and Tourette's without serious risks or enormous costs, it would have been our obligation to take advantage of it, just as it is our obligation today to provide whatever therapies they need to overcome these disabilities. The advocates of parental germinal choice and procreative liberty trust that parents will generally make the best choices for their children, and that the greatest injustice would be to insist that parents and society remain indifferent about how healthy and able-bodied our children can be.

ENHANCING THE AUTONOMY OF FUTURE GENERATIONS

Since C. S. Lewis's attack on genetic engineering in his 1965 *The Abolition of Man,* one of the dogmas of the bioLuddites, from Left to Right, is that germinal choice is somehow a tyrannical imposition on future generations.

> In reality, . . . if any one age really attains, by eugenics and scientific education, the power to make its descendants what it pleases, all men who live after it are the patients of that power. They are weaker, not stronger: for though we may have put wonderful machines in their hands we have preordained how they are to use them. . . . Man's conquest of Nature, if the dreams of some scientific planners are realized, means the rule of a few hundreds of men over billions upon billions of men. There neither is nor can be any simple increase of power on Man's side. Each new power won by man is a power over man as well.

This argument that germinal choice will rob future generations of choice can be found among all flavors of bioLuddites. In his 2003 *The Future of Human Nature,* left-wing German philosopher Jürgen Habermas tries to argue that all germinal choices must constrain egalitarian interaction simply because the parent engineered the child: "As soon as adults treat the desirable genetic traits of their descendents as a product they can shape according to a design of their own liking, they are exercising a kind of control over their genetically manipulated offspring that intervenes in another person's spontaneous relation-to-self and ethical freedom. This kind of intervention should only be exercised over things not persons." But in the course of his argument Habermas acknowledges that the genetic enhancement of children is "compatible with political liberalism only if enhancing genetic interventions neither limit the opportunities to lead an autonomous life for the person genetically treated, nor constrain the conditions for her to interact with other persons on an egalitarian basis."

Then we don't really have a problem unless we accept Habermas's tautology that all germinal choices constrain children's autonomy. Few parents intend to make their children less intelligent or less capable of autonomy and communication. If anything, parents' choices will generally expand children's ability to communicate, make decisions and control their own lives. As I say above, if there were widespread evidence that parents were systematically choosing to make their children less capable of making choices, less able-bodied, less intelligent, then I would be for regulating those bizarre choices. In fact, if anything, the problems that human enhancement may pose are from children who are far more intelligent, empowered and autonomous than their parents, leading to generation gaps.

Many decisions we make for children shape them as profoundly, and as unpredictably, as genetic decisions. Because we believe education and nutrition are essential for self-determination and success in life, we force children to go to school. No matter how aggrieved my kids feel when I force them to eat vegetables or go to school, or scream "I didn't ask to be born," it's not fascistic of my wife and I to have created them and set them on the course to healthy, autonomous adulthood. Every decision we make for our kids is a profound trust that society places in our hands in the expectation that we will act in our kids' best interests better than anybody else. If we use germinal choice to give our children more abilities, health or intelligence, we have expanded their self-determination without even the imposition on their freedom of spinach and homeroom.

A variant on the autonomy of future generations argument is that the genome should never be modified because it is "humanity's patrimony," the "common heritage of mankind." That is akin to insisting parents not paint their house, or plant a garden, or protect their trees from pests, because their children have a right to inherit the family property as is. We are constantly changing and shaping the patrimony we pass on to future generations, and it is our ethical responsibility to shape it for the better as best as we know how, and not to leave it as is.

In any case, the argument that germinal choice robs future generations of choice is also very naïve about the future course of technology. Whether genes are designed with Greg Stock's proposed on–off switches or not, enhancement technologies will evolve so rapidly that the modifications that children received prenatally will quickly be superseded by genetic and nanotechnological enhancements that can correct any flaws and inadequacies in the germinal choices made by their parents. If parents make a mistake in good faith it will likely be correctable, and our children will choose their own upgrades. If our children and grandchildren find they really needed those genes for cancer and sickle-cell anemia, they can put them back in.

CORPORATE DOMINATION AND DEMOCRATIC TECHNOLOGY

In 1999, campaigners against genetically modified food celebrated what they believed to be a major victory. The leading developer of GM crops, Monsanto, agreed not to pursue development of their so-called "Terminator gene," a genetic switch to insert in their crops to ensure that the crops produced by their seeds were sterile. The campaigners had been infuriated by the technology, since it would have required farmers to purchase the modified seeds each year. Then in 2001, researchers found strains of genetically modified corn growing in remote regions of Mexico and the alarm was raised about GM crops spreading and pushing out native crops. GM crops were also discovered to be escaping from experimental plots in the United States. Of course, one solution to the spread of GM crops in the wild would be to give them all the Terminator gene. But in the bioLuddite worldview there are no solutions that include genetic technology, with or without corporate capitalism.

Of all the bioLuddite concerns, I am most sympathetic with the complaint that profit-seeking corporations have too much control over the innovation, design, marketing and regulation of emerging technologies. But as I have said in many ways now, the solution is not to ban corporate technologies but to strengthen democratic funding of scientific research, to ensure equitable access to the fruits of technology and to provide strong independent oversight of the safety of these technologies. Corporations are driven to overhype their successes and underplay the risks of their products and services, so we need to ensure that independent clinical trials honestly represent the risks and benefits of using the drugs or therapies, and that consumers are adequately informed of those risks and benefits.

The problem of corporate priorities in the development of enhancement technologies was addressed by Pat Mooney, the executive director of the bioLuddite ETC Group, in a 2002 article in *World Watch* magazine entitled "Making Well People 'Better'." Mooney noted that "of the 1223 drugs brought to market between 1975 and 1996 only 13 targeted the deadly tropical diseases that afflict millions of the world's poor and just four of those drugs came from the private sector." Since pharmaceutical companies are profit-driven, far more resources are directed toward performance enhancement for the wealthy. Mooney then recounts with mounting disdain the enhancement drugs being developed to improve memory and control shyness, anxiety, obesity and trauma. But, to my surprise, he does not conclude that enhancement drugs should be banned, but that "if we continue to rely upon the world's giant pharmaceutical corporations to determine research goals, our societies will remain unhealthy. . . . We need to strengthen socially oriented public research and public health initiatives and, simultaneously, eliminate the patent incentive that distorts medical innovation and dictates profiteering." All of which is absolutely right, and does not require that we condemn enhancement medicine just because it doesn't cure cholera and malaria.

The struggle to work around drug patents to expand access to anti-retroviral drugs to keep HIV infections from developing into AIDS illustrates both Mooney's and my points. Anti-retrovirals are very expensive in the United States and access to them has been restricted to the world's affluent. Instead of insisting that they not be used or funded until other health needs were met, however, countries in the developing world sought the right to produce and distribute the drugs without paying fees to the patent holders that would bankrupt them. The countries would be violating the World Trade Organization's agreements on intellectual property law if they had carried through on their threat. But after a global campaign, the pharmaceutical companies reached a historic compromise in 2003 to license the drugs in the developing world much more cheaply.

It would be wonderful if the World Health Organization or the Centers for Disease Control and Prevention had developed the anti-retroviral therapies, patented them and made the patents the common property of all humankind. But so long as private sector research, motivated by profit-seeking, is the principal driver of technological innovation, the best we can do is use governments and public pressures to shape the priorities of those firms through research funding, taxation, regulation, intellectual property law, regulation on what is and isn't covered by insurance, and government purchases of drugs.

One key area in which corporations need to be carefully restrained by democracies is in the land grab of gene patenting. I argue later that genes discovered in plants, animals and humans should not be patentable, and that patents on novel gene sequences that end up as a part of conscious persons should be considered null and void. But even if we restrict gene patents, biotech firms will still have an enormous set of opportunities for profitable innovation in making the drugs and other technologies that rely on the genome.

THE VARIETIES OF BIOLUDDISM

Some of the bioLuddites I discuss here, such as the Center for Genetics and Society, are only concerned about cloning and inheritable genetic modifications and accept adults' use of nanotechnology and genetics to enhance themselves. Fukuyama and Kass want to draw a bright line between medical therapy and unacceptable "enhancement" in order to protect "human dignity." McKibben and Joy are worried about the risks to the future of the human race from new technologies, while other Luddites consider such concerns irrelevant science fiction and just want us to stop being distracted by technohype so we can focus on social reform. The Christian Right sees their crusade against transhuman technologies as a part of a broader attempt to fight sin and humanistic hubris, while Left bioLuddites believe they are fighting big corporations to defend women, the poor and the disabled.

Yet across all of these schisms the bioLuddites are finding points of convergence and principles they share in common, such as human-racism. They all share a belief that they are somehow defending humanity by denying individuals the right to use technology to make themselves and their children healthier, more able, more long-lived, smarter and happier.

Upwingers, Extropians and Transhumanists

Transhumanism, the idea of using reason to transcend the limitations of the human condition, has ancient roots. On the one hand, religious traditions have long offered ways to push back the bounds of sickness, aging and death, and achieve superhuman powers and states of mind. Millennial traditions promise a radically improved world. In transhumanism these religious aspirations have merged with humanism, rationalism, science and democracy. In the early 1990s, however, one particular brand of transhumanism mixed with Southern Californian anarcho-capitalism and became "extropianism." Although extropianism spread rapidly with the Internet, by the late 1990s transhumanism was returning to a broader, more inclusive form in the World Transhumanist Association, reembracing its democratic and humanist roots.

THE HISTORICAL ANTECEDENTS OF TRANSHUMANISM

I tell you, as long as I can conceive something better than myself I cannot be easy unless I am striving to bring it into existence or clearing the way for it. This is the law of my life. That is the working within me of Life's incessant aspiration to higher organization, wider, deeper, intense self-consciousness and clearer self-understanding.

—GEORGE BERNARD SHAW,
Man and Superman

Transhumanism is the idea that humans can use reason to transcend the limitations of the human condition. This idea has ancient roots. On the one hand, the transcendent religious traditions show that the desire to transcend sickness, aging, suffering and death is one of the most fundamental aspirations of human culture. The oldest written story of human culture, the Sumerian Epic of Gilgamesh, is about a king attempting to achieve physical immortality. Most religions offer healing, an afterlife or immortality, altered states of consciousness and a variety of superpowers—levitation, astral projection or psychic powers—to those adept at their disciplines. Many religions also promise a coming millennial paradise in which human existence will be incomparably superior.

On the other hand, reason, technology and the scientific method have been slowly extending the human competitive advantage since the invention of speech and stone tools. Rationalist humanism can be found in the earliest recorded schools of philosophy in Europe and Asia. Socrates and the sophists proposed that all human affairs were open to critical thinking, from metaphysics and ethics to the arrangement of society. The fourth-century B.C. philosopher Democritus proposed that the world known through the senses is all there is and that the world works without any prior plan. Confucius proposed codes of conduct to guide society without any referent to gods. Schools of Indian philosophy 2,500 years ago proposed that there was no afterlife and no gods, and that humans had to rely on their own reason, reflection and meditation to understand the world and be happy. The Buddhist tradition argued that each human being could achieve a superhuman state, superior to that accessible to even the gods, solely through individual human effort.

Although we think of the Dark Ages as anti-science, David Noble recounts in *The Religion of Technology* that medieval European monks developed a tradition of technology as a sacred vocation. Noble says that the Joachimite Franciscans, for instance, preached that perfecting the "useful arts" was "an approximate anticipation of, an apocalyptic sign of, and a practical preparation for the prophesied restoration of perfection." Isolated philosopher-scientists, such as the thirteenth-century Roger

Bacon, conducted experiments and kept alive the pursuit of knowledge of how things worked.

In the fourteenth and fifteenth centuries a group emerged in Europe who called themselves "the humanists." They were practicing Catholics who believed that human beings were such special creations of God that to celebrate human beings, their powers and creations was the best way to celebrate God. They condemned the theology of original sin and argued that humans should become more like God. In Italian humanist philosopher Pico della Mirandola's 1486 *Oration on the Dignity of Man,* God speaks to man saying, "to you is granted the power, contained in your intellect and judgment, to be reborn into the higher forms, the divine."

The European Renaissance and Enlightenment, from the sixteenth to eighteenth centuries, went on to forge humanism as we now know it. Renaissance humanists encouraged human beings to rely on empirical observations, reason and the scientific method, rather than religious tradition and authority. Francis Bacon's 1620 *Novum Organum* argued for the use of the scientific method to achieve a human mastery over "all things possible." Eighteenth- and nineteenth-century rationalists and free-thinkers proposed that human beings were the measure of all things. Religious humanists like Thomas Jefferson were central to the liberal democratic revolutions, creating societies that rejected divinely sanctioned rule for free association with civil liberty and separation of church and state. The French Revolution and the nineteenth-century workers movement embraced even more militantly these secular values.

The eighteenth and nineteenth centuries also saw a flowering of scientific medicine and proposals for technological means to overcome death. From Ben Franklin's wish that he could be pickled in a flask of wine and revived in a century to the immortalist speculations of Condorcet and William Godwin, democratic rationalists began to argue that social, scientific and economic progress might make possible radical life extension. Darwin's theory of evolution opened the possibility that the current condition of human beings was only a temporary stop between a prior lower and future more advanced state.

TRANSHUMANISM IN THE EARLY TWENTIETH CENTURY

The first direct intellectual precursors of transhumanism appeared in 1923 with J. B. S. Haldane's "Daedalus, Science and the Future," which proposed extrauterine gestation and genetic enhancement, and in 1929 with the Irish physicist J. D. Bernal's "The World, the Flesh and the Devil," which suggested bionic implants. Biofuturism then began spreading from Britain to the United States. In 1935, the socialist Nobel laureate biologist Herman J. Muller published *Out of the Night: A Biologist's View of the Future,* in which he predicted that humanity would soon "by aid of its ever growing intelligence and cooperation, shape itself into an increasingly sublime creation—a being beside which the mythical divinities of the past will seem more and more ridiculous, and which setting its own marvelous inner powers against the brute Goliath of the suns and the planets, challenges them to contest." Muller was also a militant opponent of racial eugenics like his contemporaries Haldane and Julian Huxley, and proposed instead that parents be able to enhance the well-being of their children through voluntary "germinal choice" (especially by availing themselves of the sperm of Nobel Prize winners like himself, but there weren't many technological alternatives at the time).

It was apparently Haldane's friend Julian Huxley, Aldous Huxley's brother, whose reflections on biofuturism led to the first use of the phrase "transhumanism." Julian Huxley argued that human beings could and should throw off the shackles of dogma and use cultural and biological means to evolve further. In his 1927 book on humanism, *Religion Without Revelation,* he wrote: "The human species can, if it wishes, transcend itself—not just sporadically, an individual here in one way, an individual there in another way, but in its entirety, as humanity. We need a name for this new belief. Perhaps *transhumanism* will serve: man remaining man, but transcending himself, by realizing new possibilities of and for his human nature."

From the 1920s to the present, speculative fiction also played an increasingly important role in stimulating biofuturist thinking by portraying both utopian and dystopian visions of human and robotic evolution. H. G. Wells and Olaf Stapledon proposed far future scenarios in which human beings subspeciated into many different forms. A science fiction subculture around the world began to grow with generally optimistic assumptions about the future, science and technology.

Secular and religious humanism also became increasingly organized and visible in the early twentieth century. *The Humanist Manifesto* was published in the United States in 1933, asserting that humanism sees "the complete realization of human personality to be the end of man's life and seeks its development and fulfillment in the here and now." Therefore humanists should "foster the creative in man and to encourage achievements that add to the satisfactions of life."

TRANSHUMANISM AFTER WORLD WAR TWO

Who are the new revolutionaries of our times? They are the geneticists, biologists, physicists, cryonologists, biotechnologists, nuclear scientists, cosmologists, astrophysicists, radio astronomers, cosmonauts, social scientists, youth corps volunteers, internationalists, humanists, science-fiction writers, normative thinkers, inventors. . . . They and others are revolutionizing the human condition in a fundamental way. Their achievements and goals go far beyond the most radical ideologies of the Old Order.

—FM–2030, *Optimism One*

Transhumanist themes exploded in speculative fiction after World War Two, exploring cloning, genetic engineering, artificial intelligence and the uplifting of animals. Starting in the early 1940s, Robert Heinlein explored the idea of a hidden subculture of people who have achieved extreme longevity through germinal choice in his Lazarus Long series. In his 1946 novel *Slan*, A. E. Vogt portrayed a future where humans violently suppress posthuman mutants. Isaac Asimov portrayed the idea of a self-aware android covertly achieving legal parity with humans in his 1950 *I, Robot*. Arthur C. Clarke pointed to the possibility of posthuman evolution in his 1953 *Childhood's End*. Even stories of human/non-human conflicts were generally optimistic that all intelligent beings could find some way to peacefully coexist, as in *Star Trek*'s Federation of Planets. In Polish writer Stanislaw Lem's 1967 *Cyberiad,* robots have long since replaced human beings.

All of these themes were also springing up outside of fiction, in social movements, medicine and computer science, feeding back into speculative fiction. In 1960, Clynes and Kline propose cyborgs to NASA. One of the goals of the program was to give ground control some way to intervene in case the isolation of space made astronauts mentally unstable. Ten years later Fred Pohl wrote *ManPlus* in which a human being is adapted to live naked on the surface of Mars, and finds he is happier as a posthuman.

As I described in Chapter 3, the cryonics movement was founded in 1962 with the publication of Robert Ettinger's *The Prospect of Immortality,* leading immediately to the founding of cryonics organizations. In 1972, Ettinger wrote *Man into Superman,* in which he made clear that the cryonics agenda was far bigger than simply getting frozen, woken and repaired when technology improves. Sex changes, redesigned digestive tracts and adaptations for extreme climates would all allow humans to transcend the limitations of the human form and a transition to "transhumanity."

In his 1968 *Toward a Psychology of Being*, transpersonal psychologist and theorist of the "peak experience" Abraham Maslow predicted the emergence of a "transhuman" psychology "centered on the cosmos" that would help human beings to become godlike. The counterculture embraced the idea that individuals could use the technology of psychedelic drugs to engineer themselves into more intelligent, integrated personalities. Although Aldous Huxley had condemned the use of the drug soma in *Brave New World* as tool of oppression, his advocacy of psychedelic liberation in the *Doors of Perception* won him a new audience with the counterculture. "Better living through chemistry" was the slogan, and when the era of psychedelics gave way to cocaine and heroin addic-

tion in the 1980s, the bio-utopians began experimenting with biofeed-
back machines and drugs for memory and intelligence enhancement.

It was while teaching at the New School for Social Research in 1966
that the Iranian-American futurist F. M. Esfandiary first used the term
"transhumanism." Like Ettinger, Esfandiary, who later changed his name
to FM–2030, used the term "transhuman" to refer to people whose
lifestyles, cultural worldviews and use of technology made them transi-
tional to posthumanity. In his 1989 book *Are You Transhuman?*, FM–2030
says transhumans "are the earliest manifestations of new evolutionary
beings. They are like those earliest hominids who many millions of years
ago came down from the trees and began to look around. . . . Many of
them are not even aware of their bridging role in evolution."

It is ironic that FM–2030 became an icon for the 1990s libertarian
strand of transhumanism since he was so clearly rooted in the New
Left's radical democratic utopianism that later birthed the Greens. Like
the Greens he argued that his politics were neither left-wing nor right-
wing, but "upwing": "The UpWing philosophy is a visionary new thrust
beyond Right and Left-wing, beyond conservative and conventional radical." But also like the Greens, FM–2030's version of transcending both
capitalism and socialism involved "the complete elimination of money
and labor." FM–2030 argued for direct electronic democracy to replace
authoritarianism and representative democracy. In place of fractious
nation-states he argued for world government and global citizenship:
"We want to help accelerate the thrust beyond nations, ethnic groups,
races to create a global consciousness, global institutions, a global language, global citizenship, global free flow of people, global commit-
ments." FM-2030's idea of Up was tilted about 45 degrees to the Left.

The women's liberation movement put germinal choice back on the
table by reintroducing the idea of human beings taking direct control of
reproduction. The feminist understanding of the scope of "reproductive rights" began with contraception and expanded to a right to abortion. But the importance of using technology to control other aspects
of reproduction was immediately obvious to many feminists, from the

nineteenth-century feminists championing anesthesia in childbirth to Shulamith Firestone's 1971 feminist classic *The Dialectic of Sex,* which argued that women cannot be finally free until they are freed from having to incubate children. Debate continues among feminists about whether plastic surgery, chemically suppressing menstruation and the use of in-vitro fertilization (IVF) are extensions of the women's empowerment over their bodies or just ways that patriarchal medicine has duped women into self-exploitation.

Many of the theologians in the emerging field of bioethics were appalled at IVF and the prospects for cloning and genetic enhancement. But some bioethicists defended the new technologies and argued that parents had rights to make germinal choices about their children, both to correct their diseases and to enhance their abilities. In his courageous and controversial 1974 book *The Ethics of Genetic Control,* University of Virginia bioethicist Joseph Fletcher made some of the first post-eugenics arguments for germinal choice. Fletcher's efficacy as an advocate was undercut, however, by a very problematic side argument for the creation of subhuman servants, which reminded many of *Brave New World.*

Much more effective, and coolly argued, was Jonathan Glover's 1984 *What Sort of People Should There Be?: Genetic Engineering, Brain Control and Their Impact on our Future World.* Glover defended human genetic engineering, virtual reality and neurotechnology, making clear that a society that permits free individual use of technologies is entirely different from the totalitarianism that people fear. Although Glover distinguished between therapy and enhancement, arguing that the former is safer than the latter, he argues that "to renounce positive genetic engineering would be to renounce any hope of fundamental improvement in what we are like. . . . Preserving the human race as it is will seem an acceptable option to all those who can watch the news on television and feel satisfied with the world."

In the late 1980s, Glover chaired a European Commission Working Party on Assisted Reproduction, which presented the landmark Glover Report to the European Commission on the *Ethics of New Reproductive*

Technologies. The Glover report was similar in tone to *Splicing Life,* a 1984 report on genetic engineering regulation produced by the bioethics commission appointed by U.S. president Jimmy Carter. Both the Glover report and *Splicing Life* defended somatic gene therapy as just another medical therapy. Neither report rejected germline therapy, but they considered it to be so difficult and dangerous as to be indefinitely off the table. The pioneering work of Fletcher, Glover and the other defenders of genetic self-determination laid the groundwork for John Robertson's articulation of the principle of "procreative liberty" in his 1994 *Children of Choice,* the idea that reproductive rights include the right to reproductive and germinal choice technology, and Julian Savulescu's articulation of the parallel principle of "procreative beneficence," that parents are obliged to choose the healthiest and most able of their possible children.

In the 1980s, computer scientists involved in artificial intelligence (AI) began to project the consequences of the creation of full, self-aware AI. Hans Moravec's 1988 *Mind Children* and 1999 *Robot* predict that robots are the next stage of evolution, and will establish a separate and incomprehensibly superior civilization. Humans will have the option of uploading their minds into these robots, "going Ex-human," but Moravec is not sanguine about the prospects for organic humanity: "Biological species almost never survive encounters with superior competitors. . . . An entity that fails to keep up with its neighbors is likely to be eaten, its space, materials, energy, and useful thoughts reorganized to serve another's goals. Such a fate may be routine for humans who dally too long on slow Earth before going Ex."

In 1986, the field of nanotechnology was born when Eric Drexler published *Engines of Creation.* In *Engines* he argued that it was not only feasible but inevitable that we will create molecular-scale robotics, capable of building anything we want from the atom up, and of reproducing themselves in great numbers. These nanobot "molecular assemblers" will make possible the complete control of the body and brain. Since Drexler was and is a cryonicist, he also made the point that nanobots will make it possible to repair the ice damage to neurons in the brains of the cryonically preserved. Drexler founded the Foresight Institute in 1986 with Christine Peterson as a vehicle to promote his vision of safe, ubiquitous molecular manufacturing and nanorobotics. Foresight has been a central node in the emerging transhumanist culture.

The first popular exposé of the emerging transhumanist culture was Ed Regis's 1990 *Great Mambo Chicken and the Transhuman Condition*. Although Regis portrayed Ettinger, Moravec and other transhumanists as humorous fringe figures, he described their ideas in thorough and accurate detail and many people first came to know about "transhumanism" through the book. But the book missed the fact that these disparate ideas had just given birth to a serious, synthesizing transhumanist movement, the "extropians."

LIBERTARIAN TRANSHUMANISM: MAX MORE AND THE EXTROPY INSTITUTE

This is really what is unique about the extropian movement: the fusion of radical technological optimism with libertarian political philosophy . . . one might call it libertarian transhumanism.

—BEN GOERTZEL,
The Path to Posthumanity

In the 1980s, a young British graduate student, Max O'Connor, became interested in futurist ideas and life extension technologies while studying philosophy and political economy at Oxford. In the mid-1980s, he became one of the pioneers of cryonics in England. After finishing at Oxford in 1988, having been impressed with the dynamism and openness to future-oriented ideas in the United States, O'Connor began his doctoral studies in philosophy at the University of Southern California. At U.S.C he began mixing with the local futurist subculture, and soon teamed up with another graduate student, T. O. Morrow, to found the journal *Extropy*.

The opposite of "entropy," the boundless expansion of "Extropy" was the core symbol for O'Connor and Morrow: life extension, the expansion of human powers and control over nature, expansion into space, and the emergence of intelligent, organic spontaneous order. O'Connor adopted the name Max More as a sign of his commitment to "what my goal is: always to improve, never to be static. I was going to get better at everything, become smarter, fitter, and healthier. It would be a constant reminder to keep moving forward."

Max More met and married FM–2030's former lover, the artist Nancie Clark, who became Natasha Vita-More. FM–2030 became a friend and supporter of the extropians, who adopted his term "transhumanism." While transhumans were an evolutionary stage for FM–2030, in the extropian lexicon "transhumanism" was a self-conscious ideological leaning. More defined transhumanism in a 1990 essay:

> *Transhumanism* is a class of philosophies that seek to guide us towards a *posthuman* condition. Transhumanism shares many elements of humanism, including a respect for reason and science, a commitment to progress, and a valuing of human (or transhuman) existence in this life rather than in some supernatural "afterlife." Transhumanism differs from humanism in recognizing and anticipating the radical alterations in the nature and possibilities of our lives resulting from various sciences and technologies such as neuroscience and neuropharmacology, life extension, nanotechnology, artificial ultraintelligence, and space habitation, combined with a rational philosophy and value system.

More always made clear that extropianism was but one of the possible varieties of transhumanism.

One way in which the extropians distinguished themselves from the broader transhumanist milieu was by making libertarianism and anarcho-capitalism central to their worldview. The young, well-educated American men attracted to the extropian milieu saw the state, and any form of egalitarianism, as a potential threat to their personal self-transformation. In the first issue of *Extropy* in 1988, More and Morrow included libertarian politics as one of the topics the magazine would promote. In early issues of *Extropy,* More began to publish successive versions and expositions of his "Extropian Principles." In the early 1990s, the Principles resolved down to five:

1. *Boundless Expansion:* Seeking more intelligence, wisdom, and effectiveness, an unlimited lifespan, and the removal of political, cultural, biological, and psychological limits to self-actualization and self-realization. Perpetually overcoming constraints on our progress and possibilities. Expanding into the universe and advancing without end.

2. *Self-Transformation:* Affirming continual psychological, intellectual, and physical self-improvement, through reason and critical thinking, personal responsibility, and experimentation. Seeking biological and neurological augmentation.

3. *Dynamic Optimism:* Positive expectations fueling dynamic action. Adopting a rational, action-based optimism, shunning both blind faith and stagnant pessimism.

4. *Intelligent Technology:* Applying science and technology creatively to transcend "natural" limits imposed by our biological heritage, culture, and environment.

5. *Spontaneous Order:* Supporting decentralized, voluntaristic social coordination processes. Fostering tolerance, diversity, foresight, personal responsibility and individual liberty.

More's fifth principle "Spontaneous Order" distilled their belief, derived from the work of Friedrich Hayek and Ayn Rand, that an anarchistic market creates free and dynamic order, while the state and its life-stealing authoritarianism is entropic. In 1991, *Extropy* magazine focused on the principle of emergent order, publishing an essay on David Friedman's anarcho-capitalist concept of "Privately Produced Law" and an article by Max More on "Order Without Orderers."

In 1991, the extropians also founded an e-mail list, catching the wind of the Internet typhoon and its high-tech libertopianism. The extremely high volume (in multiple meanings of volume) extropian list has since attracted tens of thousands of subscribers. Although there are small groups of extropians who meet together socially in Los Angeles, New York City and London, most people who consider themselves extropians have never met other extropians and participate only in this virtual community.

In 1992 More and Morrow founded the Extropy Institute, which held its first conference in 1994. At Extro 1 in Sunnyvale, California, the keynote speaker was computer scientist Hans Moravec, who repeated his cheery theme that humanity would inevitably be superseded by robots. Nanotechnologist Eric Drexler addressed the conference, and journalist Ed Regis, author of *Great Mambo Chicken,* wrote up the event for the new *Wired* magazine. Subsequent Extros, held in 1995, 1997, 1999 and 2001, have each attracted more prominent scientists, science fiction authors and futurist luminaries.

In the wake of all this attention, the extropians also began to attract withering criticism from progressives. In 1996, *Wired* contributor Paulina Borsook debated More in an online forum on the *Wired* Web site, taking him to task for selfishness, elitism and escapism, arguments she summarized in her 2001 book *Cyberselfish: A Critical Romp Through the Terribly Libertarian Culture of High Tech.* Culture critic Mark Dery excoriated the extropians and a dozen related techno-culture trends in his 1996 *Escape Velocity,* coining the dismissive phrase "body-loathing" for those, like the extropians, who want to escape from their "meat puppet" (body), those who feel a "combination of mistrust and contempt for the cumbersome flesh that accounts for the drag coefficient in technological environments."

The extropians also began to hear dissent from a growing diversity of people on their e-mail list. People sympathetic with transhumanist views but alienated by the list's abrasive, hypermasculine, libertarian politics began to amount to a sizable group. Although Natasha Vita-More has always been given prominent acknowledgment for her transhumanist arts and culture projects, and now directs the Extropy Institute, men outnumber women by at least four to one in extropian culture.

In 1997, the generally nonlibertarian European fellow-travelers of the extropians began to organize and meet under the aegis of the World Transhumanist Association (WTA). Although the WTA founders remained on good terms with the extropians, and Max More contributed to the Transhumanist Declaration and Transhumanist FAQ, the WTA's founding documents were distinctly less libertarian than the Extropian Principles.

Responding to these various trends, and presumably his own philosophical maturation, More revamped his principles in 2000 to a less libertarian Version 3.0. In this latest version More sets aside his earlier, anarcho-capitalist "Spontaneous Order" for the much more moderately libertarian "Open Society: Supporting social orders that foster freedom of speech, freedom of action, and experimentation. Opposing authoritarian social control and favoring the rule of law and decentralization of power. Preferring bargaining over battling, and exchange over compulsion. Openness to improvement rather than a static utopia." More now insists that extropianism is not libertarian and is compatible with a number of different types of liberal "open societies," just not with theocratic, authoritarian or totalitarian societies. In the extensive accompanying commentary on his new principles More even more explicitly departs from the elitist, Randian position of enlightened selfishness and argues for both a consistent rule of law and civic responsibility: "For individuals and societies to flourish, liberty must come with personal responsibility. The demand for freedom without responsibility is an adolescent's demand for license."

While More is attempting to move out of the political fringe, a casual review of the traffic on the extropian lists confirms that the majority of extropians remain staunch libertarians. In an online survey of extropians conducted in February and March of 2002, 56% of the respondents identified with "libertarian" or "anarchist/self-governance," with another 15% committed to (generally minarchist, "almost anarchy") alternative political visions. Similarly, the Extropy Institute's recommended "economics and society" readings still include David Friedman's anarcho-capitalist text *The Machinery of Freedom*, Friedrich Hayek's *The Constitution of Liberty* and the libertarian anti-environmentalist writings of Julian Simon.

As the Simon readings suggest, most extropians are ardent opponents of the environmental movement, believing that the eco-system is either not really threatened or, if it is, the only solution is more and better technology. There are occasional discussions on the extropian list about the potential downsides of emerging technologies, but these are usually waved off as being either easily remediable or unavoidable and acceptable risks given the tremendous rewards (which is not my position by the way).

The extropians have also cultivated important allies in libertarian politics such as Virginia Postrel and Ron Bailey, sympathizers with their militant defense of personal liberty and hostility to regulation and environmentalism. In 1999, Postrel, author of the technolibertarian manifesto *The Future and Its Enemies* and then editor of the libertarian magazine *Reason,* assigned *Reason*'s science correspondent Ron Bailey to focus on the defense of genetics against the Luddites. Bailey had written *ECOSCAM: The False Prophets of Ecological Apocalypse* in 1993, which argued that whatever ecological problems might have existed in the past were all being repaired now. Since focusing on biopolitics Bailey's weekly columns for *Reason* have been some of the best and most insightful critiques of the emerging Kass–Rifkin bioLuddite axis. Bailey is publishing his own technolibertarian manifesto this year, *Liberation Biology.* Postrel has now organized Bailey and other technolibertarians, such as *Tech Central Station*'s editor James Glassman and prolific InstaPundit.com writer Glenn Reynolds, into The Franklin Society. The first project of the Society has been to campaign against attempts to ban embryonic stem cell research.

UNIVERSAL IMMISERATION EXTROPIAN-STYLE

In 2003, one member of the new Franklin Society, extropian economist Robin Hanson, a professor at George Mason University, achieved his full fifteen minutes of fame. Hanson was the author of the abortive experiment by the Defense Advanced Research Projects Agency in aggregating intelligence information in a Middle East futures trading market, the so-called "TerrorDAQ." The political brouhaha around TerrorDAQ led to the cancellation of the program and the resignation of DARPA director John Poindexter.

Reporters were incredulous at Hanson and his colleagues' inability to predict how their experiment would be perceived and demagogued. While I think the experiment had merit and would not have encouraged terrorism, the episode does illustrate some of the moral and political blindness that the unreformed extropian anarcho-capitalist perspective lends itself to.

Ten years before the media lambasted Hanson for TerrorDAQ he published a now often-cited essay "If Uploads Come First—The Crack of a Future Dawn" in *Extropy* magazine. The article attempts to extrapolate the economic consequences of a breakthrough in the technology of copying human personalities into machines. He argues that the capabilities of machine-based persons would be so much greater than those of organic humans that most non-uploaded people would become unemployed. Among the uploads there would be rapid population growth and severe job competition. Eventually the enormous population of uploads would be forced to work at very low subsistence wages—the cost of their electricity and disk space—ruled over by a very few of the most successful of the uploads.

Hanson basically recapitulates Marx's vision of universal immiseration, but this time in the Matrix. In the section of the essay titled "Upload Politics" Hanson dismisses the idea that governments could impose redistribution on uploads since there would be large economic benefits of an unfettered transition to Matrix life. The average quality of life of the subsistence upload and the unemployed human would allegedly be higher than before. So the best we future residents of an uploaded society can do is become as versatile as possible to maximize our chances of ending up as one of the lucky ruling or employed classes. Hanson dismisses the idea that people will fight the division of society into a mass of well-fed plebes and a superpowerful elite since the growth in the gross domestic product is the sole measure of his utopia, and the elimination of the weak will select for "capable people willing to work for low wages, who value life even when life is hard."

With a dismal, elitist utopia like this who needs a Luddite's dystopia? Russell Jacoby has this to say of Hanson-style libertopianism: "The most imaginative futurists foresee a utopia with war, money, violence and inequality. Their future looks very much like the affluent enclaves of today, only more pleasant and commodious. They paint a picture not very different from contemporary luxury suburbs, grassy subdivisions with homes and computer and work stations set off from a larger terrain of violence and injustice. The futurists are utopians in an anti-utopian age."

THE SINGULARITY AND THE GLOBAL BRAIN

On the other hand, some extropians do have more optimistic utopian visions and expectations, or at least more optimistic versions of the liberation of human beings by machines. The idea of a "TechnoRapture," a coming utopian rupture in social life brought about by some confluence of genetic, cybernetic and nano technologies, has very old roots in the panmillennial impulse. In science fiction we have examples like Arthur C. Clarke's 1956 *The City and the Stars,* in which a paternalistic computer ensures utopia and immortality, and Heinlein's 1966 *Moon Is a Harsh Mistress,* in which an artificial intelligence helps an anarchist revolution.

Similarly the idea of a coming apocalypse is ancient and has many echoes in science fiction, from world wars, alien invasions and plagues to imperialist superintelligent machines. In the 1969 film *Colossus: The Forbin Project,* a defense department computer becomes self-aware and self-willed, takes control of nuclear weapons, hooks up with the Soviet Union's defense computer and begins issuing edicts to human beings. Again in the 1984 and 1991 films *Terminator* and *Terminator 2,* defense department computers become self-aware and decide simply to wipe out human beings.

For transhumanist millennialists and apocalyptics the seminal document is a 1993 paper on "the Singularity" by science fiction author Vernor

Vinge. Vinge projected the millennial/apocalyptic consequences of the emergence of self-willed artificial intelligence, which he estimated would occur within the next couple of decades. In physics singularities are black holes within which we can't predict how physical laws will work. In the same way, Vinge says, greater-than-human machine intelligence, multiplying exponentially, would make everything about our world unpredictable. Vinge suggests that human beings need to begin enhancing and augmenting human intelligence in order to stay one step ahead of the machines.

Transhumanist millennialism has been bolstered as a legitimate obsession for the secular technogeek by the field of nonlinear systems dynamics. Chaos and punctuated equilibrium modeling have made linear predictive models seem absurd in comparison to exponential growth models punctuated by "phase transitions" to entirely new kinds of systems with new dynamics. The concept of the sudden systemic change was popularized in Malcolm Gladwell's 2000 book *The Tipping Point,* which he defines as "that moment in an epidemic when a virus reaches critical mass, the moment on the graph when the line starts to shoot straight upwards," a phenomenon he finds in all sorts of social dynamics. Cyberpunk theorist Bruce Sterling's influential transhumanist *Schismatrix* stories in the late 1980s were inspired by Ilya Prigogine's theory that systems can abruptly self-organize into a higher level of complexity.

Not all transhumanists today believe in the immanence of a Singularity. Nor do all those who believe things will get very weird in a short space of time all focus on superintelligent machines as the *deus ex machina.* Some, like Sterling and the National Science Foundation's NBIC program, look for the seeds of exponential change in the convergence of many technologies and cultural trends. In a 2002 poll of extropians, the average year they expected "the next major breakthrough or shakeup that will radically reshape the future of humanity" was 2017, but one in five said there would be "no such event, just equal acceleration across all areas." As to the source of the next big shake-up, for those who believed a big change was coming, only a quarter of them believed it would come from artificial intelligence, while the rest believed it would be precipitated by nanotechnology and a variety of other technological and political events.

The Singularity has a special appeal for libertarians because it does not require any specific collective action. Acquiring wealth, working individually to stay on the cutting edge of technology, transforming oneself into a posthuman—these are the extropian's best insurance of surviving and prospering through the Singularity. Most Singularitarians are like pre-millennialist Christians who believed that Christians had only to prepare themselves for salvation and the millennium would be established for them, versus the "post-millennialists" who argue that Jesus will not return until the righteous turn back the tribulations and establish a kingdom of heaven on earth. TechnoRapture will elevate the techno-savvy elite who have toiled to warn and prepare the world for its coming, but found mostly derision. The unbelievers not prepared to take advantage of the TechnoRapture and be born again into new eternal bodies are likely to suffer the Tribulations of being impoverished, wiped out or enslaved. Responding to a challenge from Mark Dery about the socioeconomic implications of robotic ascension, Hans Moravec responded: "The socioeconomic implications are . . . largely irrelevant. It doesn't matter what people do, because they're going to be left behind like the second stage of a rocket. Unhappy lives, horrible deaths, and failed projects have been part of the history of life on Earth ever since there was life; what really matters in the long run is what's left over."

Again demonstrating a turn away from the libertopian fringe, extropian leader Max More has rejected the idea of a Singularity precisely because its quasi-religious millennialism contributes to passivity. According to More: "The Singularity idea has worried me for years—it's a classic religious, Christian-style, end-of-the-world concept that appeals to people in Western cultures deeply. It's also mostly nonsense. . . . The Singularity concept has all the earmarks of an idea that can lead to cultishness, and passivity. There's a tremendous amount of hard work to be done, and intellectually masturbating about a supposed Singularity is not going to get us anywhere."

There are hints of a less passive "post-millennial" or even liberation theological interpretation in some Singularitarians' work, however. For instance, extropian Eliezer Yudkowsky's central concern is building "friendliness" into the architecture of the first AIs, and convincing AI researchers to incorporate friendliness into their designs, so that the first Colossus is more Second Coming than Terminator. Although that does make friendliness research a messianic life mission for Yudkowsky, it would require collective action to be successful, and his Singularity Institute for Artificial Intelligence is devoted to the cause.

Singularitarian John Smart, director of the Institute for Accelerating Change, is more influenced by the "global brain" school of thought—that all the minds on Earth will link together through telecommunications and create a meta-intelligence. Many global brain thinkers like Smart are inspired by the writings of Jesuit paleontologist-mystic Teilhard de Chardin, who wrote in his 1940 *The Phenomenon of Man* that our increasingly dense web of thought and communication, and our technologies, are moving humanity toward integration into a "noosphere" or thoughtspace, and eventually to a spiritual telos, the "Omega Point." In 1983, New Age writer Peter Russell's *The Global Brain* suggested that the interconnection of human brains will create a global self-aware consciousness in the same way that connected neurons created human mind. Gregory Stock's 1993 *Metaman: The Merging of Humans and Machines into a Global Superorganism* provided a more materialistic and less spiritual projection of the same dynamic. The idea of the World Wide Web as the neuronal structure of a global brain quickly became a popular metaphor in the 1990s, giving extropians spending all day on e-mail and Web-surfing the sense that they were helping to build that global brain.

The global brain theorists have, in turn, converged with a school of dissident social evolutionary theorists, organized by independent scholar Howard Bloom, who argue that evolution occurs not just through the selection of individual characteristics but also through the selection of increasingly large social structures. Bloom's *Global Brain: The Evolution of Mass Mind* argues that this phenomenon would lead to global self-awareness and self-governance. Robert Wright's similar group-selectionist account of human history, *Nonzero: The Logic of Human Destiny,* suggests that world government will be an inevitable product of the trend toward larger and larger forms of cooperation and coordination.

Ironically the global brain idea has also crossbred with a central idea of the deep ecologists, the "Gaia hypothesis" pioneered in the 1980s by James Lovelock and Lynn Margulis. The Gaia hypothesis suggests that the global ecosystem has homeostatic feedback mechanisms that keep it in balance just like an organism, and ecomystics have stretched the idea to suggest that the Earth has an organismic intelligence. TechnoGaian global brainers suggest that the emerging cyborg MetaMan will incorporate or at least take collective responsibility for protecting the ecosystem, which is close to being the antithesis of the deep ecological vision.

Although the global brain, like an AI-driven Singularity, requires no specific individual or collective action, it will supposedly involve collective action. It is easy to imagine a liberation theological version of global brainism, involving social movements and organizations working toward the most liberal and egalitarian meta-human intelligence. Will the global brain have reflexes from the military-industrial complex and an appetite designed by McDonalds, or will it be built around a democratized UN and new forms of grassroots empowerment?

THE WORLD TRANSHUMANIST ASSOCIATION: LIBERAL DEMOCRATIC TRANSHUMANISM

Max More and the extropians made clear from the beginning that extropianism was but one of the many possible transhumanisms. In 1994, Anders Sandberg, the founder of the Swedish transhumanist group Aleph, noted that transhumanist ideas could be mated with many political ideologies, and that the hybrid of extropian libertarian transhumanism was just one, particularly robust, form that transhumanism could take: "Extropianism, which is a combination of transhumanist memes and libertarianism, seems to be one of the more dynamic and well-integrated systems. This has been successful, mainly because the meme has been able to organize its hosts much better than other transhumanistic meme-complexes. This has led to a certain bias among transhumanists linked to the Net towards the extropian version of the meme since it is the most widely spread and active." One European transhumanist, reviewing a conference of European transhumanists, noted: "The official program started with . . . a bleeding heart humanist socialist and a nice person. I am glad that we have that diversity among the European Transhumanists. It makes for much more refined discussions than is often seen on the Extropy mailing list."

In 1997, the Swedish philosopher Nick Bostrom (now at Oxford University) organized the World Transhumanist Association (WTA) to represent a more mature and academically respectable form of transhumanism, liberated from its "cultish" baggage. The WTA would share the techno-liberatory concerns of the extropians but allow for more political and ideological diversity. Bostrom is an academic philosopher, and the WTA project attracted several of the academics in the extropian milieu to establish *The Journal of Transhumanism* and work toward the recognition of transhumanism as a topic of academic investigation.

In 1998, Bostrom and several dozen far-flung American and European collaborators began work on the WTA's two founding documents, the Transhumanist Declaration (see Figure 10.1) and a Transhumanist Frequently Asked Questions 1.0 (FAQ1). Leading extropians, including More, contributed to the documents, but they were most heavily influenced by the Swedes Bostrom and Sandberg, the feminist and disability rights activist Kathryn Aegis and the British utilitarian thinker David Pearce. The first drafts of the documents were published in 1999.

The Transhumanist Declaration is notable in its departure from the Extropian Principles in several significant respects. Rather than calling for an unfettered technological manifest destiny, the Transhumanist Declaration specifically noted the possibility of catastrophic consequences of new technology. In the Transhumanist FAQ the authors discussed the responsibility of transhumanists to anticipate and craft public policy to

1. Humanity will be radically changed by technology in the future. We foresee the feasibility of redesigning the human condition, including such parameters as the inevitability of aging, limitations on human and artificial intellects, unchosen psychology, suffering, and our confinement to the planet earth.

2. Systematic research should be put into understanding these coming developments and their long-term consequences.

3. Transhumanists think that by being generally open and embracing of new technology we have a better chance of turning it to our advantage than if we try to ban or prohibit it.

4. Transhumanists advocate the moral right for those who so wish to use technology to extend their mental and physical capacities and to improve their control over their own lives. We seek personal growth beyond our current biological limitations.

5. In planning for the future, it is mandatory to take into account the prospect of dramatic technological progress. It would be tragic if the potential benefits failed to materialize because of ill-motivated technophobia and unnecessary prohibitions. On the other hand, it would also be tragic if intelligent life went extinct because of some disaster or war involving advanced technologies.

6. We need to create forums where people can rationally debate what needs to be done, and a social order where responsible decisions can be implemented.

7. Transhumanism advocates the well-being of all sentience (whether in artificial intellects, humans, non-human animals or possible extraterrestrial species) and encompasses many principles of modern secular humanism. Transhumanism does not support any particular party, politician or political platform.

FIGURE 10.1   The Transhumanist Declaration

prevent these catastrophic outcomes. Rather than suggesting that all social coordination can be accomplished through the market, the Transhumanist Declaration explicitly addressed the need "to create forums where people can rationally debate what needs to be done, and a social order where responsible decisions can be implemented." Here, unlike the elitist, anti-political extropians, the WTA founders took seriously the need for responsive democracies and democratic technology policies. With the Declaration transhumanists were reembracing their continuity with the Enlightenment, with democracy and humanism, and setting aside the antisocial, free-market anarchism that had briefly held sway in transhumanist circles in the unique circumstances of mid-1990s bubble economy, Southern California-based, net culture.

THE POLITICS OF THE WTA FAQ

In the last line of the Declaration, the authors make clear that the WTA is not committed to a particular political ideology. As Bostrom explained in the Transhumanist FAQ1, there are transhumanist "liberals, social democrats, libertarians, green party members." Nonetheless there are implicit political parameters to the Transhumanist Declaration and FAQ1.

The Transhumanist FAQ1 asked, "Won't new technologies only benefit the rich and powerful? What happens to the rest?" The FAQ1 acknowledged that "some technologies may cause social inequalities to widen. For example, if some form of intelligence amplification becomes available, it may at first be so expensive that only the richest can afford it. The same could happen when we learn how to genetically augment our children. Wealthy people would become smarter and make even more money."

FAQ1 argues that the answer to these inequities is not to ban the technologies, but "to increase wealth redistribution, for example by means of taxation and the provision of free services (education vouchers, IT access in public libraries, genetic enhancements covered by social security, etc.). For economical and technological progress is not a zero sum game. It's a positive sum game. It doesn't solve the old political problem of what degree of income redistribution is desirable, but it can make the pie that is to be divided enormously much greater."

Similarly, when addressing whether transhumanism is simply a distraction from the pressing problems of poverty and conflict in the world, the FAQ1 argued that transhumanists should work on both these immediate problems and futurist concerns. In fact, the FAQ1 suggests that transhuman technologies can make the solution of poverty and conflict easier, improving health care, amplifying intelligence and expanding communication and prosperity. The greatest happiness of the greatest number is a transhumanist goal in itself, and a peaceful, liberal democratic world is the best for nurturing transhuman diversity. "Working towards a world order characterized by peace, international cooperation and respect for human rights would much improve the odds that the dangerous applications of certain future technologies will not be used irresponsibly or in warfare. It would also free up resources currently spent on military armaments, and possibly channel them to improve the condition of the poor."

The FAQ1 addresses the issue of overpopulation caused by life extension technologies, arguing for both family planning and the aggressive pursuit of advanced, sustainable technologies, such as agricultural biotechnologies, cleaner industrial processes, nanotechnology and ultimately space colonization. It also notes that the best way to control population growth is to empower women since "giving people increased rational control over their lives (and especially female education and equality) causes them to have fewer children."

In response to a question about how posthumans will treat humans, the FAQ1 notes, "it could help if we continue to build stable democratic traditions and constitutions, ideally expanding the rule of law to the international plane as well as the national." Here the transhumanists are anticipating the need to build political and cultural solidarity between humans and posthumans, to minimize conflicts and to have global police institutions that can protect humans from posthumans and vice versa.

In short, the WTA documents establish a broad political tent, with an explicit embrace of political engagement, the need to defend and extend liberal democracy, and the inclusion of social democratic policy alternatives as legitimate points of discussion.

TRANSHUMANISM RELOADED

In the fall of 2001, the WTA began its next phase of growth. After publishing my paper on "The Future of Death" in the *Journal of Transhumanism,* the editor, Mark Walker, a philosopher at the University of Toronto, asked me to serve on the editorial board. Then Nick Bostrom, who was teaching at Yale at the time, Mark Walker and I organized a panel on transhumanism at the Society for the Social Studies of Science meetings in Boston in October of 2001. For that meeting I prepared a paper on "The Politics of Transhumanism" that laid out the historical circumstances that led to the accidental association of free-market libertarianism and transhumanism. Out of that meeting, Bostrom, Walker and I began to plan the expansion of the WTA.

We adopted a constitution and elected a Board of Directors. We renamed the journal the *Journal of Evolution and Technology* and launched a webzine, *Transhumanity.* I was elected secretary of the WTA in 2002. The Extropy Institute and the existing transhumanist groups in Europe became WTA affiliates, and by 2003 the WTA had two dozen mailing lists reaching thousands of people, and local groups were being organized in dozens of cities around the world. In June 2003, we hosted the first international conference on transhumanist bioethics at Yale University with sixty papers. The keynote presentations were from bioethicists Gregory Stock and Greg Pence, the libertarian science writer Ron Bailey, NSF official and co-founder of the NBIC program William Sims Bainbridge and the Yale historian of eugenics Daniel Kevles.

The WTA has been especially successful at attracting a broader swath of political views than the extropians. A membership survey conducted in December 2003, found that only a fifth of the WTAers identified with libertarianism, anarcho-capitalism or even Euro-Liberalism. On the other hand, a third of the WTA's members identified with leftist politics, ranging from labels like "libertarian socialist," "democratic socialist" and "radical" to "progressive" or "U.S.-style liberal." Conservatives of any stripe accounted for only 3–4% of members, while the plurality of WTAers were "moderate," "upwinger," "other" or "none."

TRANSHUMANISM'S INCOMPATIBILITY WITH SOCIAL CONSERVATISM

In my chart of possible ideologies in a biopolitical future (see Figure 6.3) I didn't have a letter in the corner for culturally conservative "populist transhumanists," and indeed I don't expect this ever to be a very popular variety. But that doesn't mean that there won't be some cultural conservatives, or even neofascists, who will be enthusiastic advocates of cyborgization, eugenic engineering or technotranscendence. We need only look to the origins of European fascism to see the possibility of such a phenomenon.

In 1909, the Italian writer Filippo Tommaso Marinetti published his "Manifesto of Futurism" in the Parisian newspaper *Le Figaro*. In it he called for a new aesthetic approach to life:

> We intend to exalt aggressive action, a feverish insomnia, the racer's stride, the mortal leap, the punch and the slap. . . .
>
> We want to hymn the man at the wheel, who hurls the lance of his spirit across the Earth, along the circle of its orbit. . . .
>
> We stand on the last promontory of the centuries! . . . Why should we look back, when what we want is to break down the mysterious doors of the Impossible? Time and Space died yesterday. We already live in the absolute, because we have created eternal, omnipresent speed.

Marinetti believed Europe had become stagnant and he called for a new art glorifying modern technology, energy and violence. Artists, writers, musicians, architects and many others flocked to the Futurist banner in Italy and from across Europe, and began issuing their own manifestoes. Many of the founding Futurists, including Marinetti, were anarchists. Nonetheless, most Futurists went on to urge Italy's entry into World War One, which ended the movement and its romantic calls for heroic violence and war. After the war, Marinetti befriended Mussolini, who mixed Marxist and anarchist politics with Nietzsche and heroic nationalist romanticism. Marinetti and many other Italian Futurists joined Mussolini's new fascist movement, and the fascists in turn adopted Futurist ideas and aesthetics.

Today, when a social movement emerges, such as the extropians, which scorns liberal democracy and calls for an *ubermenschlich* elite to free themselves from traditional morality, pursue boundless expansion and optimism, and create a new humanity through genetic technology and the merging of humans with machines, it is understandable that critics might associate the movement with European fascism. Nor has this problem escaped the attention of the extropians. In 1994, Anders Sandberg wrote: "many people associate ideas of superhumanity, rationally changing our biological form and speeding up the evolution of mankind, with unfashionable or disliked memes like fascism . . . partially because many transhumanist ideas had counterparts (real or apparent) among the fascists."

Extropians and transhumanists have repeatedly and forcefully insisted that transhumanism is incompatible with fascism, pointing to the transhumanists' rationalist, tolerant and libertarian values. The Transhumanist FAQ1 says:

> Racism, sexism, speciesism, belligerent nationalism and religious intolerance are unacceptable. In addition to the usual grounds for finding such practices morally objectionable, there is an additional specifically transhumanist motivation for this. In order to prepare a time when the human species may start branching out in various directions, we need to start now to strongly encourage the development of moral sentiments that are broad enough to encompass within the sphere of moral concern sentiences that are different from current selves.

In March of 2002, as its first official position statements after the adoption of FAQ1, the World Transhumanist Association voted to formally denounce "Any and all doctrines of racial or ethnic supremacy/inferiority [as] incompatible with the fundamental tolerance and humanist roots of transhumanism." The strong transhumanist condemnation of racialism appears to have succeeded in dissuading racialist groups from trying to join or recruit among transhumanists.

The anti-racist implications of widespread germinal choice technology have also given racists pause. In a thread on the neo-Nazi Stormfront Web site titled "Is Transhumanism Good for White Nationalists?" one poster notes:

> What's wrong with this form of egalitarianism? After all, if everyone is genetically engineered with superior intelligence, blacks, whites, yellows and all, then the world would be a much better place. The problem with egalitarianism today is that people are trying to make equal that which is simply not equal. But if everyone were truly equal, there would be no need to make everyone equal, and therefore no need for egalitarianism.

But another poster objects:

> I have some real concern about the ability of the White race to use these technologies wisely in the present situation. Eugenics in recent decades has largely meant going to a sperm bank to have the child of a Jewish medical student.

TRANSHUMANISM AND DEMOCRACY

Speaking to the Extro 5 conference in 2001, extropian leader Greg Burch argued that transhumanists were culturally and politically encircled by religious fundamentalists, Greens and socialists: "We do not seek to force our plans on anyone, but ultimately, our basic values of individual autonomy are fundamentally incompatible with the kinds of limitations desired by Guardians of both culturally conservative and 'progressive' tendencies."

The transhumanist perspective is indeed outnumbered by much better organized and more influential opponents, and libertarian extropians like Burch are partly to blame. The ideological narrowness and sectarianism of the briefly ascendant libertarian transhumanists of the 1990s are striking in comparison to the ideologically diverse coalitions forged by the bioLuddites. While Burch and the extropians argue that they are fighting to save the Enlightenment, in fact they are fighting to extol only one-third of the Enlightenment—liberty—to the exclusion of the other two-thirds—equality and solidarity. In the process they have crippled their ability to defend all three values. Insisting that reason can only be expressed in market relations and not in rational civic debate and democratic self-governance leaves the anarcho-capitalist transhumanists self-absorbed and alienated from serious political engagement, unable to respond to either the public's legitimate or illegitimate anxieties about the future.

On the other hand, a much broader spectrum of thought is expressed by the World Transhumanist Association, more reflective of the political and cultural diversity of the popular transhumanist constituency waiting to be heard. As twenty-first-century biopolitics matures, the WTA and the term "transhumanism" may not be important players in the struggle for individual rights to use human enhancement technologies. But the shift of the transhumanist subculture away from libertopianism to movement-building and serious engagement with public policy is a hopeful sign that transhumanists may be able to play an important role.

FREEDOM AND EQUALITY AMONG THE CYBORGS

Democratic Transhumanism

The democratic humanism of the French and American revolutions has inspired dozens of movements, all united by the idea that humans should use reason and democracy to control their own lives. In this chapter I argue that those diverse threads can be united in a radically democratic form of techno-optimism, a democratic transhumanism. Democratic transhumanism is the next stage of human self-emancipation through science and democracy. Democratic transhumanism addresses the legitimate concerns of the bioLuddites for equity, solidarity and public safety, and libertarian concerns with our right to control our bodies and minds. If libertarians want enhancement technologies to be safe, widely available and unhampered by Luddite bans, they need to support legitimate regulation and universal provision. If progressives want enhancement technologies to make society more equal, they need to make enhancement universally available. Numerous constituencies and movements can be woven together into a democratic transhumanist politics, including advocates of reproductive rights, disability rights, universal basic income, drug decriminalization and transgender rights.

MARQUIS DE CONDORCET, DEMOCRATIC TRANSHUMANIST

One of the most interesting early champions of democratic transhumanism was the French philosopher and revolutionary Marquis de Condorcet. When Marie Jean Antoine Nicolas Caritat Condorcet was elected to the French Academy of Science at age 26, he had already demonstrated a broad and eclectic scientific curiosity. Condorcet went on to contribute fundamental insights into the mathematics involved in elections and voting. An aristocrat, friend of Voltaire and member of the *philosophe* school, his lectures helped popularize the sciences to a growing middle-class audience. He also was a passionate revolutionary and campaigned actively for political freedom, religious tolerance and the abolition of slavery. After the French Revolution, Condorcet was elected to the French Legislative Assembly, where he worked to establish public education in France. His draft constitution for the French guaranteed broad personal liberties. But when it fell out of favor with the Jacobins he and his associates were sentenced to the guillotine.

Condorcet spent his final months hiding in Paris working on *A Sketch for a Historical Picture of the Progress of the Human Mind*. The *Sketch* was intended to be the introduction to a longer history of the effect of science on humanity. But Condorcet never had a chance to write that larger work. In March 1774, Condorcet finished the *Sketch*, left his hiding place and was promptly arrested. He died shortly thereafter, perhaps from suicide or simply from exhaustion, before he could be executed.

The *Sketch* is a remarkable document for its optimistic faith in humanity's ability to liberate itself with technology and democracy. It is doubly remarkable that it was written under sentence of death from a regime that allegedly represented those very principles. But Condorcet's was a utopian vision that looked beyond the Jacobin terror as a mere detour in the course of history. In the *Sketch* he argues that human beings are using reason and science to free themselves from domination by one another and Nature. He argues that "nature has set no term to the perfection of human faculties; that the perfectibility of man is truly indefinite; and that the progress of this perfectibility, from now onwards independent of any power that might wish to halt it, has no other limit than the duration of the globe upon which nature has cast us."

Technology and democracy, according to Condorcet, were both the result of the natural process of reason and intelligence growing and asserting itself. Reason throws off the shackles of prejudice, tyranny, elitism and ignorance. Social science, a form of applied reason, makes possible the more perfect social order of liberal democracy, which, in turn, supports the continuous freeing of the human mind.

> The time will therefore come when the sun will shine only on free men who know no other master but their reason; when tyrants and slaves, priests and their stupid or hypocritical instruments will exist only in works of history or on the stage; and when we shall think of them only . . . to learn how to recognize and so to destroy by force of reason, the first seeds of tyranny and superstition, should they ever dare to reappear amongst us.

Reason demands not just equality for all men, including non-Europeans, but also gender equality. Sexism holds back the "progress of the human mind that is most important for human happiness," not just for women, but "even to the sex it favors." Patriarchy has "no other source but the abuse of power."

Condorcet wanted universal public education, especially in the sciences. He wanted a system of social insurance to provide for senior citizens, widows and eventually everyone. Giving everyone a basic income would be inevitable since human beings would be freed from labor by automation and agricultural methods that produce more from each plot of land. Equality, education, preventive health care and better food would all extend longevity, till we conquered death altogether.

> Would it be absurd now to suppose that the improvement of the human race should be regarded as capable of unlimited progress? That a time will come when death would result only from extraordinary accidents or the more and more gradual wearing out of vitality, and that, finally, the duration of the average interval between birth and wearing out has itself no specific limit whatsoever? No doubt man will not become immortal, but cannot the span constantly increase between the

moment he begins to live and the time when naturally, without illness or accident, he finds life a burden?

Even under sentence of death, Condorcet saw the French Revolution as the beginning of humanity's liberation from human oppression and the oppression of nature. In 1972, Henry Kissinger asked Chinese prime minister Chou En-lai what he thought of the French Revolution. After a long pause Chou replied, "It's too early to tell." Indeed, the repercussions are still being felt.

SOCIAL DEMOCRACY

Democracy is not a state, it's a direction. No society is anywhere close to an ideal democracy. Since the French and American revolutions there have been many variants, combinations and interpretations of liberty, equality and solidarity, ranging from Wild West free-market minimalism to authoritarian populism. Generally the problems of these models can be framed as an inadequate respect for one or another of the values of liberty, equality or solidarity. Immature authoritarian democracies give too much weight to racial, religious and nationalist solidarity and not enough to liberty and equality. Populist and communist regimes focus on equality to the exclusion of liberty. The religious Right, from Islamic fundamentalism to the Bush administration, has the most restrictive interpretation of democracy, opposing all expansions of liberty, equality or solidarity. In the Wild West corner the libertarian tradition seeks to expand personal and economic liberty, but cares little about their effect on equality and solidarity. For the libertarians all efforts to help the poor and powerless should be voluntary since taxes are state-sponsored theft.

Empirically the best current balance of liberty, equality and solidarity is found in social democratic Canada and Europe, where all three values have been maximized. Canadians and Europeans have personal liberties, such as gay marriage and drug decriminalization, that American libertarians can only dream about, and at the same time more egalitarian distributions of wealth, more leisure time and more generous social welfare systems.

The degree of equality in a country has been summed up by a single statistic, the Gini coefficient, reported by the World Bank. The Gini coefficient varies between 0.0, or perfect equality, and 1.0, where one person has everything and everybody else has nothing. On this measure the most equal of the industrial societies are former communist Eastern Europe and social democratic Western Europe, with Gini scores between 20 and 30. The United States ranks seventy-third (Gini score = 40) out of the 118 countries that the World Bank reports on, a littler more unequal than Ghana and a little more equal than Senegal.

In terms of liberty, the United Nations produces a Human Freedom Index, which summarizes

- the right to travel, assemble and speak
- the absence of forced labor, torture and other extreme legal punishment (such as the death penalty)
- freedom of political opposition, the press and trade unions
- an independent judiciary
- gender equality
- the legal right to trial, counsel of choice, privacy, religion and sexual practice.

The UN Human Freedom Index ranges from 0, for the least free countries, to 40 for the most free. In 2002, the United States ranked tenth in freedom, below social democratic Europe: Sweden, Denmark, the Netherlands, Austria, Finland, France, Germany, Canada and Switzerland.

As a consequence of being more free, more equal and having more generous welfare policies, the social democracies also consistently rank as having the highest quality of life in the world. For instance, every year the United Nations produces its Human Development Index, which scores each country based on its average life expectancy, average income per person and average educational attainment. In the 2002 report the United States doesn't do that badly, since our high adjusted average income compensates for our lower life expectancy and educational attainment. But the top five countries in the index are social democratic Norway, Sweden, Canada, Belgium and Australia.

Richard Estes at the University of Pennsylvania produces an even more ambitious index of "social progress" measuring forty variables, including health, education, human rights, political participation, gender equality, cultural diversity and social stability. By this measure the "social progress" of the United States has stagnated since 1980, ranking now with Poland, while the top ten countries are again in social democratic Europe: Denmark, Sweden, Norway, Finland, Luxembourg, Germany, Austria, Iceland, Italy and Belgium.

All this suggests a linear scale of democratic progress, from the plutocratic and authoritarian democracies of the former communist and developing world, to free-market liberal democracies like the United States, to the European social democracies. But even Canada and Europe are still just imperfect approximations of the more ideal democracy that we are capable of, a democracy in which we are all freer, more equal, more informed and more involved. European and Canadian citizens can become better educated and participate more actively in policy-formation, and wealth and corporations could have even less influence in shaping public opinion and public policy there as here. Even the European democracies are forced to bend a knee to the demands of international capital. But for much of the world, including the United States, achieving a European level of prosperity and social democracy will be an enormous achievement, one that will take many more decades.

This chapter argues for a social democratic or "radical democratic" version of transhumanism, a future-positive, techno-optimist democracy. This "democratic transhumanism" is framed by its debate with its two closest competitors, the Left bioLuddites and the libertarian transhumanists.

WHY DEMOCRATS SHOULD REEMBRACE TRANSHUMANISM

We want nothing less than the right to determine our own evolution. We want the right to live forever—to succeed with our revolution against death itself. So long as we have not overthrown the tyranny of death, all mankind belongs to the developing world, all mankind is proletarian.

—FM−2030, *Optimism One*

I wrote part of this book in a beautiful wood-lined room in a commune-run retreat in western Massachusetts. The commune advocates politically progressive, ecofriendly mysticism, a deep ecological suspicion of modern technology mixed with paganism. My family and I were attending a Buddhist family retreat there, and the experience was like Zen comes to Hobbiton. While we watched our breath in meditation in their beautiful, sunny, all-wood meeting hall, the residents and staff practiced organic gardening techniques with hoes and scythes in the scattered vegetable gardens, producing all the food consumed in our vegetarian meals. Toilets had two seats, one for solid waste and one for urine, which were disposed of in ecofriendly ways involving sawdust. The beautiful post-and-beam buildings had small octagonal windows and thick walls stuffed with cotton matting made from ground-up blue jeans. The effect was lovely, and I think many people could enjoy living this kind of lifestyle if given the choice. Certainly many tourists with comfortable middle-class homes to return to, like me, enjoy living that way for a weekend.

The problem for progressive politics is that living this kind of lifestyle permanently is not attractive for the vast majority of the world still trapped in *involuntary* simplicity. Most people do not want to live in a future without telecommunications, flush toilets, labor-saving devices, air travel and medicine. Most people don't have those things now, and suspect their lives would be a lot better if they did have them. When given a choice between the spiritually rich, politically correct, voluntary simplicity advocated in *Utne Reader* and the technoporn fantasy future found in *Wired,* most of the world utters a couple rote words of scorn for decadent Western techno-affluence and then desperately tries to figure out how to climb into the pages of *Wired.*

Luddism is a political dead-end for progressive politics. Left-wing Luddism is boring and depressing, and has no energy to inspire people to create a new and better society. The Left was built by people inspired by millennial visions, not by people who saw only a hopeless future of futile existential protest against the juggernaut of fascist Progress. If there is to be a future for progressive politics it has to come from a rebirth of a sexy, high-tech vision of a radically democratic future, a rediscovery of the utopian imagination. As Russell Jacoby says in *The End of Utopia,* "in an era of political resignation and fatigue the utopian spirit remains more necessary than ever. It evokes neither prisons nor programs, but an idea of human solidarity and happiness. . . . Something is missing. A light has gone out. The world stripped of anticipation turns cold and grey." What is missing, the light that has gone out for the Left, is the idea that the human condition can be radically transformed, that we can accomplish more than a defense of the status quo against a capitalist version of the future. To rekindle a progressive utopianism, the Next Left, the twenty-first-century Left, needs visionary projects worthy of a united transhuman world, projects like guaranteeing health, intelligence and longevity for all, building world government, eliminating work and colonizing the Solar System.

Luddism is also bad political sociology. Left Luddites inappropriately equate technologies with the power relations around those technologies, and try to fight capitalism or patriarchy or hierarchy by fighting technologies instead of by liberating the technologies for free and equal use. Technologies may make certain kinds of power more likely than others, but they do not *determine* power relations. Each new technology creates a new terrain for organizing and democratic struggle, new possibilities for expanded liberty and equality, or for oppression and exploitation. Technological innovation needs to be democratically regulated and guided, not fought or forbidden.

Progressives need to reembrace the Enlightenment insight that liberating each individual's potential requires not only political liberation, but also technological liberation from nature. Marx referred to technological progress as the move from the realm of necessity to the realm of freedom. The more powerful our technology, and our political and economic empowerment, the more we are freed from the necessity of labor.

Democratic transhumanism combines this old strain of progressive optimism about reason, science and technology with a strong defense of individual liberty. The assertion of individual liberty was also a central cause of the radical democrats, and only became identified with the libertarian Right because of the rise of communism. By embracing the right of each person to control their body and mind and freely use technology to realize their fullest potential, the Next Left can decisively break all associations with authoritarianism.

A political movement based on both technological progress and individual liberty will then see ways that democratically regulated and distributed, freely exercised technology can create a more equal, empowered and united world. One way is by reducing the biological bases of social inequality. Contrary to the vacuous assertions of Francis Fukuyama and Bill McKibben that we are all biological equals, a lot of social inequality is built on a biological foundation, and enhancement technology makes it possible to redress that source of inequality. FM–2030 wrote in *Optimism One* that transhumans

> are no longer content simply striving for social, economic, and political equality. What do these rights mean so long as people are *born* biologically unequal? So long as some are born strong others weak, some healthy others sickly, some beautiful others ungainly, some tall others short, some brilliant others dumb—in other words so long as we do not have biological equality—all social equalities mean very little. We will settle for nothing less than [the conquest of] this basic biological inequality which is at the very root of all human inequalities.

Patriarchy, the most fundamental form of human domination, begins with the fact that men can beat women up, and that women are doubly vulnerable when pregnant. Gender equality does not depend on reengineering women to have more upper-body strength, and there are examples of pre-industrial societies with pretty good gender equality. But modern technology—health care and birth control that gave women decades free from children, factories that reduced the economic significance of male strength, the telephone and police car and pistol that provide protection from male violence—makes it possible to create a post-patriarchal society. Of course all of these technologies can also be used by men in the service of male power, but their overall effect in democratic societies has been liberatory.

In the coming decades reproductive technology will make it possible to liberate women even more radically from the dictates of nature. Birth control, abortion, menstruation-regulating drugs and cesarean section have already given women control over their wombs. Feminist utopian novelists were depicting female-only societies that reproduce through parthenogenesis as early as Charlotte Perkins Gilman's 1915 *Herland*. The next step, as New Left feminist writer Shulamith Firestone argued in the 1971 classic *The Dialectic of Sex,* is the artificial womb, which will give women complete choice. A character in Marge Piercy's 1976 novel *Woman on the Edge of Time* explains why her future utopian society uses artificial wombs: "It was part of women's long revolution. As long as we were biologically enchained, we'd never be equal. And males never would be humanized to be loving and tender. So we all became mothers."

Beyond gender, biological differences in abilities also create social inequality. Innate intellectual endowments obviously do predict, even if they don't determine, life success, as do many other biological traits. Short men earn less than tall men. Obesity, which is substantially genetic, is a strong predictor of the length and success of our lives. People with chronic physical and mental illnesses are dependent on others and not able to accomplish as much, despite the accomplishments of the extraordinary few. And what is more disempowering than dying young? To the extent that we make technologies for greater health, ability, longevity, intelligence and happiness universally available, those technologies will make it possible for all biologically disadvantaged people to have more real equal opportunities in life. As Nicholas Wade says in *Life Script,* genomics makes it possible "to envisage for the first time the creation of a genetically more just society, one in which the most fundamental kind of wealth—the genes that confer health and fitness—would for the first time be accessible to all."

Recognizing these biological sources of inequality does not mean accepting the moral legitimacy of inequality, it just means opening up a second, complementary front in the struggle against those inequalities. Of course we still need to work to overcome exploitation, domination and prejudice at the same time that we use technologies to make them harder to sustain. When parents are able to choose the height of their children, more children will be tall, but we should also work on being less heightist. There is no contradiction between thoroughly attacking our classist stigmatizing of fat people and at the same time giving them the technologies they need to achieve whatever body image they prefer, even if their ideal bodies are drawn from rail-thin magazine models. There is no contradiction between promoting generous social supports and workplace accommodations for the mentally disabled and promoting technologies that cure or prevent mental disability.

The egalitarian theorist best known for this argument is Ronald Dworkin, who argues for a regime of liberal, egalitarian eugenics in his book *Sovereign Virtue.* Dworkin starts with two principles. First, he argues that a person's life, once begun, should succeed rather than fail, "that the potential of that life be realized rather than wasted." Second, he argues that each person has the right to define for themselves what a successful life is, which is a restatement of John Stuart Mills' rationale for liberty—we know our own needs better than anybody else does. Adopting those two principles, Dworkin argues, leads inescapably to a society that provides universal access to germinal choices and human enhancement technology. Universally available enhancements will make it most likely that each person will reach their highest potential. This principle is actually already written into the Universal Declaration on the Human Genome and Human Rights, adopted by the UN in 1998: "Benefits from advances in biology, genetics and medicine, concerning the human genome, shall be made available to all."

Another theorist who argues that egalitarians should embrace subsidized germinal choice technology, including enhancements, is the Princeton University bioethicist Peter Singer. In Singer's 2001 *A Darwinian Left: Politics, Evolution, and Cooperation,* he argues that the Left has ignored and denied the sociobiological constraints on politics to its own detriment. Singer contends that there is a biologically rooted tendency toward selfishness and hierarchy in human nature that undermines egalitarian social reforms. If ambitious egalitarian programs of social reform and democratic cooperation are to succeed, Singer argues, we must employ the new genetic and neurological sciences to identify and modify the aspects of human nature that cause conflict and competition. "In a more distant future we can still barely glimpse, it may turn out to be the prerequisite for a new kind of freedom: the freedom to shape our genes so that instead of living in societies constrained by our evolutionary origins, we can build the kind of society we judge best." Toward that end Singer advocates a program of voluntary, socially subsidized genetic enhancement.

We all may need genetic and cybernetic enhancement to finally satisfy the demands of engaged citizenship. Biological limitations don't only constrain the capacity for democratic self-governance of the cognitively disabled, but of all humans. FM–2030 (echoing the New Left) wrote that transhumans want "instant universal participation that will do away with the very institution of government." But direct participatory democracy, as in an Athens or the New England town meeting, requires a high level of information collection, organizing, reasoning, argument and activism that is beyond the capacities of most people. Once a community is larger than 2,500 people or so the issues become so complex that we begin to delegate to more or less accountable elites. Few of us have the leisure, education and interest to make educated interventions about a wide range of state, national and global issues. Even a clearly posthuman intellect like Noam Chomsky doesn't write much outside of foreign policy and linguistics.

As George Bernard Shaw remarks in the *Revolutionist's Handbook*, "Democracy cannot rise above the level of the human material of which its voters are made. . . . [Democracies will continue to be swayed by demagogues] unless we can have a Democracy of Supermen; and the production of such a Democracy is the only change that is now hopeful enough to nerve us to the effort that Revolution demands." Human enhancement technologies promise to expand our capacity for citizenship, making direct, participatory, electronically mediated democracy more possible. Our future brains, wired to the world through telecommunications, will be capable of both thinking and acting globally. We will monitor the world with special expert systems and make political decisions based on more sophisticated heuristics than a politician's party affiliation or religious views.

Already by answering a couple dozen questions in an online survey at SmartSelect.com you can determine with high accuracy your affinity for political ideologies and religious beliefs that you may never have heard of. The computer scientist Jason Tester has designed an expert system dubbed "Constituty" that will monitor your e-mail and the Web pages you read to make guesses about your political ideology, and the candidates and issues you are likely to support. The system could then track how well candidates fulfill campaign promises, and reward them with your additional support. Once we have expanded our capacities for knowledge, attention, deliberation and communication, even a small proportion of our energies may be enough to read journals, monitor C-SPAN, participate in online debates and vote on the UN referenda, while the rest of our brain gets on with the more important things in our lives.

Expanding individuals' capacity to participate and make informed choices does not guarantee equality, of course, since those already wealthy and endowed will use their advantages to accumulate even more information and secure their interests. But just as the literate, well-fed citizens of the 1960s insisted on forms of democracy undreamt of by eighteenth-century sharecroppers, increasing the health, intelligence, longevity, education and leisure of the ordinary citizen will make them more capable of recognizing how an unequal society does not serve their interests, and more able to understand the methods they need to pursue to achieve empowerment.

Jonathan Glover noted in *What Sort of People Should There Be?* that it "is not just any aspect of present human nature that is worth preserving. Rather it is especially those features which contribute to self-development and self-expression, to certain kinds of relationships, and to the development of our consciousness and understanding. And some of these features may be extended rather than threatened by technology." More subtle than the biological constraints on equality and self-empowerment are our social and psychological constraints, the unequally distributed brainwashing and neuroses that rob us of our fullest potential. With nano-neuro technology we will have unimaginable access to and control over our currently unconscious reactions to the world, our ingrained deference to hierarchy, our addictions and self-destructive behaviors, and the ways that we are manipulated by advertising, charismatic authority and social approval. As we liberate and unkink our personalities we will also be liberating the world.

A final reason for democrats to embrace transhuman enhancement is that it may be the only way to keep liberal and social democracies competitive with authoritarian regimes. Since communism collapsed, when few expected it to, we've been struggling to find the new political paradigm for the New World Order. While liberal and social democracies have an enormous advantage they are by no means assured of long-term supremacy. If the democracies hobble themselves with restrictions on human enhancement and technological innovation, we can be sure that the authoritarian regimes will not.

Of course I'm thinking of China. Although I am still confident that China will become an open democracy in a decade or two, this optimism may be as misplaced as our previous certainty about the longevity of Russian communism. The booming prosperity of Chinese authoritarian capitalism has dried up sympathy for Tiananmen Square-style democratic revolution, and the Chinese have refocused on leveraging new technologies to their advantage. China has the only explicitly eugenicist laws in the world, and beyond a ban on reproductive cloning the Chinese have few ethical quandaries about stem cell cloning or plant, animal or human genetic engineering. In 2002, 46% of Chinese students graduated with engineering degrees compared to 5% in the United States. More than 600 nanotechnology firms have started in China. The Chinese space program plans a base on the Moon in the next decade.

The Chinese government has scaled back its ideological and hegemonic ambitions. But there is still a real possibility of conflict between the growing bloc of liberal and social democracies, laboring under self-imposed bans on human enhancement and emergent technologies, and militarily and economically robust authoritarian regimes like China using human enhancement, nanotechnology and AI to full advantage. While we press to create global governance capable of effectively controlling technological threats to global security, the democracies need to retain the technological advantage. If the transhuman technological advantages are as sudden and dramatic as I believe they will be, our long-term choice may be transhuman democracy or no democracy at all.

WHY LIBERTARIANS SHOULD EMBRACE TRANSHUMAN DEMOCRACY

Social democracy is not simply the ideal balance of liberty with equality and solidarity, but the most effective known *guarantor* of liberty. As the Nobel Prize–winning economist Amartya Sen and Ronald Dworkin have argued, social democracy provides real freedom for real people, people who need access to health care and universal education in order to fully participate in society and function as citizens. Worldwide, and even within the industrial democracies, poor and working-class people are less able to access the education, leisure, health care and wealth to express themselves and give their kids the same opportunities in life as the afflu-ent. Creating a truly free world, a world in which individuals will be best able to realize their aspirations, requires making a more equal world.

Libertarians and egalitarian democrats have fundamentally different premises about the nature of freedom, however. Libertarians want the government not to interfere with the individual's right to make decisions in free markets. The egalitarian democrat wants everyone not only to be free to make fully informed decisions, but also to have the means to make those decisions. The libertarian agrees with the egalitarian that everyone should have an "equal opportunity" in life, but disagrees that redistribution or social services are required in order to guarantee equal opportunity. The egalitarian thinks that democracy is required to con-trol for the tendency of power and resources to accumulate to people with advantages. The libertarian thinks that markets level out advan-tages more effectively than governments.

Given these fundamental differences in values and worldviews I have not had much success using this or other arguments in convincing well-educated libertarians of the virtues of social democracy. For every study I produce showing that a social program is effective, or that social democracies are more healthy, prosperous and free than the United States, they can produce a right-wing think tank's study that says the op-posite. Even when there is tangible evidence of a "market failure" the libertarian always believes the answer is to apply more market mecha-nisms instead of better democratic governance. Measured against the imaginary perfect free market, actual democracies always lose.

So instead of making libertarians into social democrats I want to build a case for democratic transhumanism for the technolibertarian who ac-cepts that there is some minimal rationale for the existence of govern-ment. If we can agree that governments can do at least a couple things better than markets, we may be able to map out why various democratic policies may be necessary to bring more of all good things to all people.

It has always struck me as very odd, though, the way most transhu-manist libertarians argue for the market. Citing libertarian guru

Friedrich Hayek, they insist that the market is a naturally evolved, emergent phenomenon without conscious guidance, which allocates resources better than any conscious human planning ever can. That is a strange argument from people who believe that all of the imperfect products of evolution can and should be redesigned by human beings. Most libertarians accept that human beings are smart enough to control the genome and the ecosystem better than nature, but balk at the idea that we could ever improve the economy with planning or regulation. Many nineteenth-century humanists became socialists precisely because they believed the market could be improved by rational human design, which has certainly proved to be the case even if it cannot yet be replaced altogether.

Of course the idea that the market is more "natural" than political institutions is just as silly and unfounded as the idea that selfishness is more natural than altruism, or that women are more in touch with nature than men. As Gus DeZerega argues in his Hayek-inspired defense of democracy, *Power, Persuasion and Polity,* markets and democracies are both products of evolution and conscious human action. Just like evolving markets, democracies require millions of autonomous agents aggregating their interests, expressing themselves in competition and cooperation. Like markets, democracies are subject to monopoly and information failures. Markets, in turn, require laws, legislatures, police, courts, planning and the provision of public goods like education, health care, defense and roads. Both democracies and markets can be improved through the application of human reason, planning and regulation. Markets and democracies depend on one another and any ideal society we can conceive of needs a combination of both.

Anyway, the most powerful arguments for the need for government, and one that many libertarians grudgingly accept, are threats to the existence of the human race. Most libertarian transhumanists grudgingly acknowledge that nanotechnology, genetic engineering and artificial intelligence could cause catastrophes if used for terrorist or military purposes, or if accidentally allowed to reproduce in the wild, and that governments can play a role in reducing these risks. Transhumanists who seriously want to be around in a couple hundred years see the need for governments to monitor, regulate and prepare for these risks.

Of course, some optimistic libertarians propose that we don't really need governments to regulate and prepare for these risks. We could force firms to take fewer risks by holding them accountable through product liability. The free market could provide cheap space suits and nanobot-off spray at Wal-Mart, and we could form voluntary clubs to contract with private mercenaries to take out nano-terrorists. So, the apocalyptic risks argument for government only works for those who feel safer threatening North Korea with the International Atomic Energy Agency and UN sanctions rather than with Rambo and lawsuits, which fortunately includes most libertarians.

Believing that one will be around indefinitely also significantly increases the chance that you will be on the planet when the next asteroid hits, or gamma-ray burst sterilizes the local star belt. If it is possible to survive these threats, the solutions again lie, at least in part, in government action. The Fortune 500 aren't going to set up asteroid monitoring networks, or send ICBMs to blast asteroids, even if it would be good for their bottom line. Getting human beings off the planet so that we don't have all our eggs in the third basket from the Sun is also not going to be accomplished through market mechanisms, which is why the most serious libertarian space advocates have grudgingly accepted the need for government-financed space programs. Securing the "existential benefits" of the new technologies requires as much governance as preventing their existential risks.

Short of threats to the future of life on Earth, most libertarians can also agree that laws should require the full disclosure of relevant information when purchasing products or making contracts. It is not much of a step from the need for consumers to have full information, or "fully informed consent" in bioethics terms, to the requirement that new products be independently tested for safety and efficacy, and that those tests should be shared with consumers. Libertarians can even agree that corporations should not be able to use their disproportionate power in society to undermine the testing of their products or suppress results.

Disagreement comes in the government's responsibility to protect people from risks they have been clearly warned about. While most people think the government has some obligation to ban unsafe products and chemicals, libertarians will argue that the government should restrict itself to a "buyers-beware" *Consumer Reports* function. But ensuring that all emerging nano- and biotechnologies are adequately tested by independent agencies, and that those results are made public, would still be a huge advance over the current haphazard and spottily effective regulatory system.

Libertarians will also embrace regulation of neurotechnologies because of the threat they pose to our understanding and experience of individual freedom. Although Fukuyama and Kass's attacks on Prozac and Ritalin are overwrought, the use of future neurotechnology or genetic engineering to produce obedience in children, soldiers or citizens, even if accomplished through free choices, could be such a threat to the liberty of the next generation that we would need to stop it by law. Imagine, for instance, that we achieve the drug legalization that I and most transhumanists advocate, and then a drug is developed that—unlike current drugs—is 100% and permanently addictive. The drug might rewrite the brain so that all goals and values become secondary to remaining intoxicated. Suppressing or discouraging such a drug would be an exercise of coercion in the defense of liberty, keeping people from selling themselves into slavery.

Getting serious about risks to the public and proposing nonmarket solutions for those risks is the only way to fight the bioLuddite agenda politically. Only believable and effective policies that guarantee technologies are safe and equitably distributed can reassure skittish publics. Panglossian assurances that all will work itself out in the market or after the Singularity won't cut it. We will face much more opposition to enhancement and radical life extension if they are only available to the rich. Without universal health access and economic security, the shrinking working-age population, fewer and fewer of whom are able to afford health insurance, will wage war against the "geezers." If we don't promise serious answers to structural unemployment—expanding the welfare state, a guaranteed basic income, expanded access to higher education, job retraining, a shorter workweek and worklife— then we are likely to see the return of old-school Luddite machine-smashing by the unemployed. National health insurance and a basic guaranteed income also provide more choices, of physician and occupation, than private health insurance or the naked free market. Public policies can address and ameliorate the public's legitimate concerns, slowing innovation in the short term but facilitating innovation and choice in the long term.

Another area where libertarians are increasingly sounding like socialists is in the critique of effects of rampant intellectual property claims. People who are serious about seeing technologies rapidly developed and made widely accessible to the public at reasonable prices have to be concerned that patents are strong enough to encourage innovation, but not so strong that they suppress competition and make technologies too expensive. Many libertarians are coming to the conclusion that the current overly expansive intellectual property system is holding technological innovation back.

Finally, my appeal to libertarians returns to the personhood project that is the common root of libertarian and radical democratic politics, and of humanism and transhumanism. Protecting the rights and freedoms of individuals so that they can fulfill their potentials means we have to create cultures and governments that tolerate and protect diversity. Securing and protecting rights for posthuman intelligence will require alliances with movements for sexual, reproductive, cultural, racial and religious equality. Libertarian transhumanists bring to the table legitimate objections to bureaucratic methods to enforce group rights, but they must not let these qualms keep them from working in solidarity with transhumanism's natural allies for a tolerant and diverse transhuman democracy.

WEAVING A NEW DEMOCRATIC TRANSHUMANISM

Free in Body and Mind

At the 2003 Transvision conference at Yale University, Vanessa Foster, the chair of the National Transgender Acton Coalition, took the podium in "The Future of Sex and Gender" workshop and announced that transsexuals like herself were the first transhumanists. It was an important moment in the history of transhuman politics. Transhumanism as a vanguard civil rights movement had arrived, and the stunned but open expressions on the faces of the largely straight male audience showed the work that transhumanists still needed to do to reach out to the disparate constituencies that will build a democratic transhumanism.

First among these constituencies are the disparate movements, like transgender rights, working to radicalize our control over our own bodies. Reproductive rights activists, who insist that women should have subsidized access to reproductive and contraceptive technology, are natural allies of a democratic transhumanism. Although many feminists have been influenced by ecofeminist bioLuddism and Left Luddite arguments about the danger of corporate technology, there is a broader feminist constituency that sees no contradiction between women's empowerment and using technology to expand their control over their lives, so long as those technologies are safe.

A transhumanist ideological thread that has grown in academia for the last twenty years is found in the cyborgology of Donna Haraway. In 1984, Donna Haraway wrote "A Manifesto for Cyborgs: Science, Technology, and Socialist Feminism in the 1980s" as a critique of Luddite ecofeminism, and it landed with the reverberating bang of a hand grenade. Haraway argued that it was precisely in the eroding boundary between human beings and machines, in the integration of women and machines in particular, that we can find liberation from patriarchy and capitalism. Haraway says, "I would rather be a cyborg than a goddess," and proposes the cyborg as the liberatory mythos for all women. Haraway's essay and subsequent writings have inspired the new subdiscipline of "cyborgology" or "cyberfeminism." As yet there has been little cross-pollination between the left-wing academic cyborgologists and the transhumanists, but mutual recognition and ties are growing.

Gays, lesbians and bisexuals are also natural allies of democratic transhumanism since the right to control one's own body means being able to share it with other consenting adults. The champions of natural law attack homosexuality and human enhancement with the same arguments. Invitro fertilization allows lesbians to have children without having sex with a man. Work on fertilizing eggs with the DNA from another egg or replacing egg DNA with sperm DNA would allow gay parents to both have a genetic link to their children.

One activist who saw that link and ran with it is veteran gay rights activist Randy Wicker. Wicker was one of the first gay rights campaigners to come out on radio and television in the early 1960s, and he has been active in gay rights in New York City ever since. Then in 1996, when an international backlash started against the cloning of the sheep Dolly in Scotland, Wicker had an epiphany. He saw that the right to clone was a fundamental reproductive rights and gay rights issue since "cloning renders heterosexuality's historic monopoly on reproduction obsolete." Wicker started the Clone Rights United Front with other gay rights activists, then co-founded the Human Cloning Foundation, and has become a national spokesman on cloning as a reproductive right. Wicker's argument has made some headway with gay theoreticians like Chandler Burr, author of *A Separate Creation: The Search for the Biological Origin of Sexual Orientation,* who acknowledges that reproductive technology "takes us another degree further from the idea that babies are produced only by two heterosexual people having heterosexual intercourse."

Another enormous constituency for democratic transhumanism are the tens of millions of people whose lives are harmed by laws against cognitive liberty, that is, laws against illicit drugs. Drug dependency is a huge public health problem and should be treated as such. Yet all the problems associated with drug use are made worse by the Drug War. In this century drugs and other brain control technologies will only become more complex, and the technologies of surveillance and repression more powerful. A society that denies us the right to put cannabis in our brain, and forces us to pee on demand to prove we haven't, is a society more likely to tell us we can't use intelligence enhancers and mood modifiers and more willing to use new technologies of repression to ensure we don't. For instance, drug vaccines that prevent the action of specific drugs are not simply being developed as voluntary tools for people trying to kick addictions, but as preventive measures that businesses can require their employees to take.

A far better use of public monies, as Aldous Huxley proposed in *Doors of Perception* and transhumanist David Pearce proposes in "The Hedonistic Imperative," would be to develop better drugs with fewer health risks. Opposing the ill-conceived and fruitless Drug War puts democratic transhumanists in solidarity not only with the millions of political prisoners serving time for nonviolent drug use and possession, but also with the cutting-edge activists for cognitive liberty, such as Wrye Sententia and her Center for Cognitive Liberty and Ethics, who is working to "establish, promote, and protect the right of each individual to use the full spectrum of his or her mind, to engage in multiple modes of thought, and to experience alternative states of consciousness."

Disabled Cyborgs and Secular Scientists

Disabled people using the latest assistive technologies, with their eyes fixed on medical progress, are also a natural constituency for democratic transhumanism. Disabled people in the wealthier industrialized countries, with their wheelchairs, prosthetic limbs, novel computing interfaces and portable computing, are the most technologically dependent humans ever known, and are aggressive in their insistence on their social rights to inclusion, support and treatment. Some disabled people are even consciously embracing the transgressive image of the cyborg. In an article in *Wired,* paraplegic journalist John Hockenberry made the point that disabled people are pushing the boundaries of humanness: "Humanity's specs are back on the drawing board, thanks to some unlikely designers, and the disabled have a serious advantage in this conversation.

They've been using technology in collaborative, intimate ways for years—to move, to communicate, to interact with the world. . . . People with disabilities—who for much of human history died or were left to die—are now, due to medical technology, living full lives. As they do, the definition of humanness has begun to widen."

Probably the most prominent symbol of disabled transhumanist activism these days is Christopher Reeve, the former Superman actor who became a tireless campaigner for biomedical research after a horse-riding accident left him quadriplegic. Reeve has been especially eloquent defending the use of cloned embryos in stem cell research, and his advocacy of cures for spinal injuries has made him controversial in the disability rights community, some of whom see a zero-sum trade-off between disability rights and cures.

But most disabled people are not Luddites. Most disabled think we can allow parents to choose to have nondisabled children and that technology can be used to overcome or cure disabilities, while we fight for equality for people with disabilities. Just as we should have the choice to get rid of a disability, we should also have the right to choose not to be "fixed," and to choose to live with bodies that aren't "normal." The right not to be coerced by society to adopt a "normal" body is also a central demand of transhumanism.

Patient advocacy groups and scientific lobbies also share a broad interest with the transhumanist movement in seeing more public financing of medical research and protecting the freedom to conduct research from bioLuddite bans. The struggle against the anti-science policies of the Bush administration and the Republican Congress, from bans on stem cell research to suppression of climate change research, has mobilized an enormous coalition in Washington, D.C., in defense of scientific research. The Coalition for the Advancement of Medical Research, the principal pro-stem cell research lobby, includes patient groups like the American Diabetes Association and the American Infertility Association, physician organizations like the American Medical Association and the American College of Obstetricians and Gynecologists, research universities such as the University of California System, education associations such as the American Council on Education, foundations like the one founded by Christopher Reeve, and industry groups like the Biotechnology Industry Association and National Venture Capital Association. This polarization of the scientific, medical and academic communities against the Luddism of the Christian Right is very good news for the emergence of a progressive transhumanist political movement.

Scientists in particular are a democratic transhumanist constituency. Most American scientists are secular, civil libertarian and lean toward the Democrats. Scientists believe passionately in scientific freedom, are incredulous at neoLuddite attacks on technological progress and are suspicious of the religious fundamentalist base in the Republican Party. The Republicans noticed. Bush's powerful political advisor Karl Rove told the *New Yorker* that the definition of a Democrat was "somebody with a doctorate." Bush has half as many Ph.D.s in his cabinet as Clinton did, and he moved the Office of Science and Technology Policy outside the White House and cut its staff.

Scientists have grown even more restive as the Bush administration dismissed the scientific consensus on stem cells, climate change, Head Start and abstinence-oriented sex education. In February of 2004, a thousand scientists signed on to a Union of Concerned Scientists report condemning the Bush administration's suppression and distortion of science. When the stacked conservative majority on the President's Council on Bioethics recommended the banning of therapeutic stem cell cloning, every practicing scientist voted against the resolution. So in March of 2004 the administration replaced two pro-stem cell members of the Bioethics Council, one of them a scientist, with three Christian bioLuddites. Almost two hundred bioethicists signed a letter of protest.

Biopunks and Open-Source Science

As I mentioned earlier, libertarians have begun to join progressives and many scientists in a revolt against the quickly expanding intellectual property laws in biotech that are retarding scientific and technological progress. Science writer Annalee Newitz has connected this revolt against the closure of the bio-commons to an emerging "biopunk" ethos in the work of biotech-inspired artists. Biopunks, according to Newitz, are committed both to the benefits that can emerge from genetic technology and to opposing the madness of patents on discovered genomes that allow corporate control of genetic data that should be in the public domain. Biopunks protest both "bioLuddites and apologists for the biotech industry."

Newitz finds biopunk sensibilities expressed in groups like the Coalition of Artists and Life Forms (CALF), a loose network of artists who celebrate biotechnology while remaining critical of its capitalist exploitation and limitations. Biopunk sensibilities among scientists, Newitz argues, can be seen in the growing call for the "open sourcing" of scientific information, from free scientific journals to public domain experiments like the GenBank and Gene Expression Omnibus database, and the International Consortium on Brain Mapping.

Technogaians and Viridians

Democratic transhumanists need to make the case to people concerned about public and ecosystem health that new technologies can be developed safely and deployed in ways that prevent and repair the damage humans have done so far. Political philosopher Walter Truett Anderson is an example of a democratic transhumanist technogaian who has tried to make that case. In *To Govern Evolution* and *Evolution Isn't What it Used to Be*, Anderson argued that humans need to take seriously our democratic responsibility for managing nature, both in the ecosystem and in our genome: "Today the driving force in evolution is human intelligence. . . . Even our own genetic future is in our hands, guided not by Darwinian abstractions but by science and medical technology and public policy. . . . This is the project of the coming era: to create a social and political order— a global one—commensurate to human power in nature. The project requires a shift from evolutionary meddling to evolutionary governance."

Technogaianism applied to ecosystem management is found in the "reconciliation ecology" movement and writings such as Michael Rosenzweig's *Win-Win Ecology*. Rosenzweig boils down his approach to several simple steps: "First, drink deeply from the natural history of the species you want to help. Study their reproductive cycles, their diets, and their behavior. Abstract the essence of their needs from what you observe. Then apply it without worrying whether your redesign of the human landscape will resemble a wilderness. It won't, so feel free to be outrageously creative."

One of the most outrageously creative of technogaian thinkers is the science fiction author and cyberpunk ideologist Bruce Sterling. In January 2000, Sterling returned to his polemicist roots and penned a 4,300-word manifesto for a new "Viridian" green political movement. Sterling accepts the urgency of climate change and species depletion. But he believes Green politics are too Luddite and dour. He calls for a sexy, high-tech, design movement to make attractive, practical ecological tools. Like FM–2030, Sterling outlines a third way between capitalism and socialism involving controls on transnational capital, redirecting militaries to peacekeeping, developing sustainable industries, increasing leisure time, guaranteeing a social wage, expanding global public health and promoting gender equity. The Viridian movement has attracted hundreds of people who receive weekly missives from Sterling about exciting ecologically appropriate technologies.

Many researchers in biotech also have ecological motivations. The AgBioWorld Foundation at the Tuskegee Institute has mobilized a global network of biotech scientists to defend biotech crops on humanitarian and ecological grounds. For instance, crops can be genetically engineered that require less agricultural land, pesticides and fertilizer, and provide more essential nutrients. In a review published in 2004, the UN Food and Agriculture Organization stated that the principal problem with biotech crops is that small farmers have not been able to adopt them more quickly. The report concluded that there have been no environmental or health harms from GM crops, and many benefits such as reduced use of pesticides and increased productivity.

Nanoecologists have also begun to emerge to propose a wide range of ecological applications of nanotechnology, such as Eric Drexler and Chris Peterson in *Unbounding the Future* and Doug Mulhall in *Our Molecular Future*. In its 2003 review of nanotech and AI titled "Future Technologies,

Today's Choices," Greenpeace says there is no need for bans on nanotech, or even new regulatory structures, and that "new technologies . . . are also an integral part of our solutions to environmental problems, including renewable energy technologies such as solar, wind and wave power, and waste treatment technologies such as mechanical-biological treatment."

While transhumanists and deep ecologists profoundly disagree about population control (transhumanists want more people, and deep ecologists want fewer), we agree on the goal: a high-tech society with individual rights and a strong democratic state that will cause people to have fewer children. Industrializing liberal democracies reduce childhood mortality and provide women with education, employment opportunities and publicly financed family planning, giving them the means and incentives to control their fertility. Affluent liberal democracies ensure the well-being of the elderly through social security systems so they don't need large numbers of children as their de facto pension plan.

And Everybody Else . . .

As machine intelligence becomes more sophisticated and automates all manual, service and intellectual labor, we may all need a pension. Since the original Luddite revolt techno-utopians have predicted the end of labor and the spread of leisure, and to a degree they were right. At the turn of the century most working people in the industrialized countries worked 3,000 hours a year from their early teens to the day they died. Today we work about 1,600–1,900 hours a year, and the average person lives ten years into retirement. In the coming decades the work life will continue to shrink, and we may already be seeing the beginning of the true end of work in the weak recovery that began in 2003 with jobs being lost to automation and outsourcing as corporate profits grew.

The job-loss recovery was predicted by Hans Moravec in his book *Robot*, where he also says society won't put up with growing inequality and concentration of wealth: "It is unlikely that a future majority of service-providing 'commoners' with more free time, communications and democracy than today would tolerate being lorded over by a dynasty of non-working hereditary capitalists. They would vote to change the system. The trend in the social democracies has been to equalize income by raising the standards of the poorest as high as the economy can bear. In the age of robots, that minimum will be very high."

Moravec then suggests that capitalism will come to an end and that society will need to provide a universal basic income: "Incremental expansion of such a subsidy would let money from robot industries, collected as corporate taxes, be returned to the general population as pension payments. By gradually lowering the retirement age, most of the population would eventually be supported . . . [and] payments begun at birth would subsidize a long, comfortable retirement for the entire original-model human race."

In a similar vein, Marshall Brain, the computer scientist and entrepreneur who founded the successful HowStuffWorks Web site and book series, is promoting his "Robot Nation" epiphany, that half of all jobs in the United States will be lost to the developing world or robots by 2055. Brain suggests that all Americans should receive a guaranteed basic income of $25,000 a year, paid from a general fund supported by progressive taxation, corporate fines and the sale of public resources. Brain argues that a guaranteed basic income is necessary for the survival of capitalism: Capitalism may be able to do without workers, but not without consumers.

Moravec and Brain join a growing international movement of economists and activists advocating a "basic income guarantee" (BIG). BIG is the answer to the next wave of Luddite machine-wrecking by angry displaced workers. The Luddites have no faith that democracy can allow everyone to benefit from technological innovation, and the libertopians think we don't need democracy and public assistance since we have the stock market. But Brain, Moravec and the BIG movement aim to prove that democracies can provide universal economic benefits while advancing the technological innovation necessary to pay for them.

Universal health care and basic income systems are essential as we make the transhuman transition, to ensure equal access to benefits not only between the rich and poor, but also between the young and old. As the population rapidly ages, and the population supporting senior benefits shrinks both demographically and because of structural unemployment, generational conflict will be inevitable without programs that provide universal benefits. Either the shrinking population of angry young workers will wage war on the benefits available to the growing senior and unemployed population, or we will expand the benefits of income security and health insurance to everyone.

A QUICK REVIEW

Let's quickly review the lay of the ideological land. On matters of personhood-based citizenship, natural law, individual rights and the "yuck factor," libertarian and democratic transhumanists are on the same page. But on questions of regulating technology, the democratic transhumanists side with most of the rest of the world against the libertarian and Luddite extremes: appropriately regulate technology and avoid bans. On issues ranging from social equality and joblessness to ecological protection and globalization, the libertarians argue that the market will solve all problems while the democratic transhumanists argue for judicious government regulation and universal social provision. These issues and positions are presented in Table 11.1.

BUILDING A DEMOCRATIC TRANSHUMANIST MAJORITY

Currently all the self-described "democratic transhumanists" in the world could hold a convention in a large classroom. That's not the point. There is a latent majority constituency for social justice, a caring society, technological progress and health and longevity for all. Even though no politician would get elected yet on a platform of ape rights, subsidized intelligence enhancements and a universal guaranteed income, the basic goals of democratic transhumanism are shared by the vast majority of people. The challenge is to find issues and struggles

TABLE 11.1 Biopolitical Ideologies

	Libertarian Transhumanists	Democratic Transhumanists	Left-Wing BioLuddites	Natural Law BioLuddites
Citizenship	**Personhood-Based "Cyborg Citizenship":** All self-aware beings with desires and plans for the future should be considered citizens with a right to life		**Limited Human-Racism:** "Humanness" determines citizenship, but not for embryos	**(Religious) Human-Racism:** "Humanness" determines citizenship **(Deep Ecological) Eco-centrism:** Human beings have equal rights with all living things
Humanism vs. Natural Law	**Humanism:** Human beings are free to determine their own future, guided by prudent reason. There are no obvious natural or divine limits on human aspiration		**Opposition to (Capitalist/ Imperialist/Patriarchal) Humanism**	**Deep Ecology/Hubris Taboo:** Humanity be restricted by should divine or ecological taboos
Individual Liberty vs. "Yuck Factor"	**Individual liberty** trumps "yuck factor"		**"Yuck factor"** trumps individual liberty in germinal choice and biotech, but not in sexuality or abortion	**"Yuck factor"** trumps individual liberty
Technological Risks	**Resignation:** Technology is uncontrollable,	**Regulation:** Risks are manageable with the assistance of democratic oversight and management	**Relinquishment:** Risks are so enormous and unknowable, and regulatory institutions so flawed, that human enhancement should be banned	

(continues)

TABLE 11.1 Biopolitical Ideologies

The Equality Challenge of Enhancement Technology	**Market Access and Legal Equality Enough:** If legal equality is guaranteed and enhancement technologies are available in the market, social equality is irrelevant and government should do nothing to create a more equal society	**Make Enhancement Universally Accessible:** Democracies should work toward social equality and provide universal access to enhancement technologies	**Bans on Technologies as Part of Larger Egalitarian Program:** Democracies should work toward social equality and ban enhancement technologies	**Tech Bans Necessary to Protect Equality:** Equality can be guaranteed by banning enhancement technologies
Procreative Liberty	**Procreative Liberty:** Trust parents to act in kids' best interests, and let them buy germinal choice in the market	**Procreative Liberty, Equality and Beneficence:** Trust parents to act in kids' best interests, stop them if they don't, give them equal access to germinal choice technology and encourage them to create children with the best life opportunities	**Limited Procreative Liberty:** Reproductive rights, but not extended to germinal choice and reproductive technology	**Procreative Liberty Trumped by Natural Law:** Reproductive rights trumped by religious prohibitions or the need for radical population reduction
Ecological Protection	**Free Market Green:** The market can solve all ecological problems	**Technogaian:** A combination of judicious regulation and ecologically oriented technologies can prevent and remediate ecological damage	**Left Ecology:** Technologies developed by patriarchal, capitalist systems need to be banned and those systems changed	**Deep Ecology:** Humanity should radically restrict itself **(Religious) Dominion Theology:** Man has stewardship over Nature

(continues)

TABLE 11.1 Biopolitical Ideologies

Structural Unemployment	**The Market Will Provide:** If the government avoids meddling (unemployment insurance, minimum wages, etc.) all workers will find new jobs, even if at lower wages	**Embrace the End of Work:** The wealth and leisure created by automation should be shared equitably by all through a basic income guarantee and shorter work week	**Protectionism, Luddism, Public Employment:** Attempt to protect existing jobs, and create new ones	**(EcoLuddite) Small Is Beautiful:** Stop automation and adopt voluntary simplicity
Globalization	**Economic Globalization Good:** Global governance, worker protections, environmental laws, all unnecessary	**Economic Globalization Must Be Accompanied by Political Globalization:** Economic globalization good so long as it is accompanied by worker rights and protections, environmental laws and global democratic governance of capital flows		**(EcoLuddite) Globalization Should Be Stopped:** Return to local self-reliance and political sovereignty **(Religious) Global Bans on Enhancement Tech:** Global treaties to ban cloning and enhancement, but no other global governance

that coalesce that latent majority and demonstrate how marginal the libertarians and bioLuddites really are.

In 1996 the National Opinion Research Center asked a random sample of Americans whether it was the government's responsibility to provide health care for the sick. As has been true since the first time that survey question was asked in the 1930s, a dominant majority, 85%, said yes. The survey also asked whether genetic screening was likely to produce more harm or more good, and two-thirds thought it would produce more good than harm (Table 11.2). Using just those two questions, left-leaning techno-optimists constitute more than half of all Americans, or 56%. That is the majority waiting to be having its voice heard.

TABLE 11.2   The Latest Democratic Transhumanist Majority

		Do you think genetic screening will do more good than harm or more harm than good?	
		More Good	More Harm
Is it the government's responsibility	Yes	56%	27%
to provide health care for the sick?	No	12%	5%

Source: General Social Survey 1996, National Opinion Research Center (N=311).

Transhuman Democracy

In this chapter I sketch out some of the policy ramifications of a democratic transhumanism. The cyborg citizenship approach suggests which kinds of beings should be granted citizenship and a right to life, and which should be considered property. It also leads to some surprising conclusions about our obligations to the great apes. Next, I show that the idea that persons own themselves has some surprising consequences for the patenting of human genes. Finally, I suggest a method for deciding which enhancements should be made universally available, and which should be discouraged or banned, all while maximizing individual health care choices.

TYPES OF PERSONS AND WHAT WE OWE THEM

As I suggested in Chapter 7 on cyborg citizenship, a central question of biopolitics will be what rights we should grant to the various kinds of beings we create with technology. The human-racists want to restrict rights to *Homo sapiens* 1.0, while the transhumanists, like many bioethicists and the democratic tradition itself, believe rights should be based on personhood.

But the argument quickly becomes more complex since there is no agreement about what personhood entails. Whatever list of characteristics one uses, technology is sure to create cases that lie exactly in the gray areas. The most minimal threshold for personhood is something like "conscious beings, aware of themselves, with intents and purposes over time." In addition to consciousness, self-awareness and self-motivated activity, philosopher Mary Anne Warren adds "reasoning" and "the capacity to communicate." Other personhood theorists, such as Tristram Englehardt, Joseph Fletcher and Joel Feinberg, also add the ability to care about the opinion of others, which I discuss later.

All personhood theorists agree, however, that the basic threshold of citizenship is self-awareness and desire. Once you are self-aware and looking forward to your plans for the future, even if those plans only concern your next meal, your death is qualitatively more significant, at least to you. You are a being in yourself and not just a thing. Once a being achieves self-awareness, therefore, is when it makes sense to say that the being is a person with a "right to life," and cannot be owned as a thing.

Incomplete persons should not be considered property, nor can we treat them yet as full citizens. Children are self-aware beings with plans and desires, but it makes no sense to grant them the right to control their own lives, to make contracts or vote, until they have matured into their full adult powers. As children they still have poorly formed "intents and purposes over time," insufficient ability to reason about their own or other people's interests, and immature capacities for empathy and communication. If we granted citizenship rights to great apes, the only meaningful rights they could possess would be the right to life, the right not to be bought or sold, and freedom from being subject to medical experimentation. The right to vote, make contracts or get a college loan would be meaningless for a chimpanzee. At least until she is genetically enhanced or "uplifted," which I'll address in a second.

So a personhood approach to citizenship, "cyborg citizenship," recognizes two different statuses of citizenship, disabled and full. The most intelligent animals, such as great apes and dolphins, can be included in our polity with the limited rights that we grant to human children, the demented, the mentally ill and the severely mentally disabled. These are disabled citizens, since their ability to fully determine their own affairs and participate in democratic deliberation is compromised. Full citizens are persons who fulfill all of Warren's conditions such as adult humans, posthumans (at least the kind that most of us would choose to become), uplifted animals and possibly some future human-level machine minds.

Animals, humans or machines that lack self-awareness would not be citizens, not even disabled citizens, and therefore are not candidates for citizen-level rights. Things that are not citizens are necessarily property, and their affairs are determined by laws controlling what people can do with different kinds of property. Not all property can be bought or sold, and you can't always dispose of property any way you like. Even though you can't sell your brain-dead relative's organs, their body is still essentially family property that you can decide how to dispose of within the framework of the law.

But here we need to make an additional distinction between sentient property, property that can feel but is not self-aware, and nonsentient property (Table 12.1). Things that can feel pleasure and pain oblige us to ensure that they not be caused unnecessary suffering. Sentient property includes fetuses, most animals and permanently unconscious humans who are still sensate. Simple biological property, which can't feel pain, would include things like embryos and corpses.

These categories also inform our obligations to beings. We are obliged to help fellow citizens achieve their fullest capacities for reason, consciousness and self-determination. If it is possible to enable disabled citizens to achieve a fuller possession of their faculties for reason, autonomy and communication, so that they can control their own affairs, it is our obligation to help them do so. When our fellow citizens are less able, less healthy, less intelligent or less happy than they otherwise could be, it is our ethical and political responsibility to do what we can, while respecting whatever self-determination they are able to exercise.

Caring is sometimes all we can do. But quite often, as a society, we can do quite a bit more. We can assure adequate nutrition and clean

TABLE 12.1   The Continuum of Consciousness and Rights

Example Types of Life	Consciousness	Rights Status
Adult Humans, Enhanced or Unenhanced, and Their Cognitive Equals	Mature Personhood, with Reason	Full Citizens *Right and Ability to Self-Determination, to Vote and to Make Contracts*
Human Children Demented and Mentally Disabled Human Adults Great Apes	Personhood (Self-awareness)	Disabled Citizens *Right to Life and to Assistance to Achieve Full Self-Determination*
Most Animals Fetuses Permanently Vegetative Humans	Sentience (Pleasure and Pain)	Sentient Property *Right Not to Suffer Unnecessarily*
Brain-Dead Humans Embryos Plants Toasters	Not Sentient	Property

water for everyone on the planet. We can make sure that people are not killed or crippled by war, landmines and toxic wastes, and that they are not raped and tortured by brutal regimes. We have an obligation to children to provide them with education and secure homes so they can realize their abilities. We have an obligation to the mentally ill to provide them with treatments that return them to sanity.

Alongside the provision of basic needs, education and a caring community, we also are increasingly able to offer technology as a means for people to reach their fullest potentials. Assistive technology for the disabled is the most dramatic case of the technological empowerment of citizens. For instance, in the rare neurological condition of "the locked-in state," the peripheral nerves of the body stop working but the brain is unaffected. As many as 25,000 Americans are estimated to suffer this level of severe or total paralysis. Slowly the victim becomes encased in a body with which they can't communicate. They can hear and see, and they are just as conscious and awake as ever inside, but they can't even twitch an eyelid. As patients slip into this horrifying state they have often been counseled to take their own lives or request that no measures be made to keep them alive. But now researchers at Emory University have put special computer chips into the brains of patients in the locked-in state, in the area that once controlled their muscles. The patients' neurons grow into contact with the chip and when the patients think about moving various muscles radio signals are sent to an external computer. Slowly the patterns of neuronal firing against the chips' electrodes allow the patients to control computer cursors by thought alone. Totally paralyzed people are now surfing the Web, sending e-mail and controlling their wheelchairs and their lives by thought alone.

Future stem cell and neural regeneration techniques will allow people with spinal cord injuries to walk and control their bodies. Then it will soon be possible to give great apes and dolphins genetic and cybernetic enhancements of their intelligence and language skills to help them reason and communicate.

Did I lose you there? It seems bizarre to most that we might have any obligation to intelligent animals beyond leaving them alone in the wild, and treating those in captivity humanely. But I think we have the same obligation to uplift "disabled" animal citizens that we have to disabled human citizens. This does not mean that we are obliged to round up all apes or dolphins in the wild and give them gene therapy, brain drugs and education. Capturing apes and dolphins violates the autonomy they currently exercise, and would mean far more death and suffering than uplift. Probably the most meaningful thing we can presently do to promote the rights of dolphins and apes in the wild is to protect them from hunters and fishing nets, and their habitats from destruction. As to apes and dolphins in captivity, we need to ensure that they are treated with respect, and not subjected to medical experimentation since they are unable to consent to participate. As technologies progress, however, it will become increasingly practical to provide dolphins and chimpanzees in captivity with technologies that allow them to think more complicated thoughts and communicate with humans, ranging from systems that translate between human speech and animal thoughts to genetic enhancements of their brains.

Any program of uplift for intelligent animals, children or disabled humans unable to consent needs to be conducted very carefully in case their enhancement causes greater suffering. Even if we believe with John Stuart Mill that it is better to be an unhappy human than a satisfied pig, we should try to avoid making intelligent but miserable pigs. We might find, for instance, that tweaking the intelligence of animals increases their sensitivity to pain or their neuroticism. If there were such downsides to the upgrades we should probably hold off, just as we would rethink sending children to schools if they all became literate but miserable adults.

Our obligation to cybernetic intelligence is much more complicated and bounded by caution, since organic people will likely face more significant threats from machine minds that achieve self-awareness than they do from enhanced chimpanzees, *Planet of the Apes* notwithstanding. Machine minds are far less certain of having the capacities for empathy and morality critical for full citizenship, and their potential to harm human interests will increase much more rapidly than organically embodied intelligence. Our obligation to acknowledge self-aware machines will need to be balanced by our obligation to protect the interests of already existing organic citizens. Ensuring that machine minds either do not achieve self-awareness, or achieve only a safe level of powers with inbuilt solidarity for the rest of their fellow citizens, will likely require a pervasive set of as yet unimagined regulations and policing practices.

So the model of cyborg citizenship I've outlined here makes clear which kinds of creatures we have obligations to. We may feel a fondness for trees and rivers, and we may want to preserve them and the animals that live in them, but they do not bear rights. We do not have an obligation to uplift beings that are not self-aware, who have not achieved the threshold of citizenship, such as fetuses or fish. Fetuses and fish, like all matter in the universe, are potentially self-aware, but we are not obliged to make all matter in the universe intelligent.

OWNING OUR BODIES

If you paint a picture of a mountain, you aren't granted ownership of the mountain. You aren't even given the right to forbid other people from painting pictures of the mountain. Even if you spend ten years climbing all over the mountain to make a detailed, computerized topographic map that tells you where every rock came from and where every river runs, you still don't get a deed to the mountain or a patent on the idea of mapping that mountain. You can, however, get a copyright on your painting or map, and if someone else wants a picture or map of the mountain they have to either buy it from you, buy someone else's painting or map, or make one of their own. If you invented a new mapping method, or a new kind of ink, you can get patents on those, and people will have to pay you to use them to map the mountain. But whoever owns the mountain still can tell you to get off her land.

This seems straightforward, and yet it is not at all obvious to the U. S. Patent and Trademark Office, and the patent offices of many other countries, who have granted tens of thousands of patents on the use of human and animal genes to companies that only decoded them. Since patents are only supposed to be granted for *inventions* that are useful, nonobvious and *new,* biotech companies have argued that the decoded gene is a new useful invention, even though that information is there for anyone to discover. The patent offices, under tremendous pressure from the biotech industry, accepted this argument to facilitate biotech innovation. The biotech firms argued that they had no incentive to invest in making a truly novel drug or gene therapy based on genetic discoveries if they couldn't have exclusive rights to that slice of the gene.

Understandably gene patents have outraged not only bioLuddites, but scientists worried about research access to patented genes and public health activists worried that gene patents will make gene-derived medicines prohibitively expensive. Transhumanists also have a huge stake in this debate since the genetic therapies we hope to have widely available are both being motivated and restricted by this set of property rights. If gene patents were necessary to spur useful innovation, there might be a pragmatic reason for their use, but they appear to be having the opposite effect, and be unnecessary to provide incentives for drug development. For instance, all the anti-cholesterol "statins"—Lipitor, Zocor, Pravachol, made by three different firms—target the same liver enzyme. Yet they all work differently, and one drug, Baycol, had to be taken off the market because of side effects. If one firm had been able to get a patent on the gene for that enzyme, the patent-holder would have prevented this very useful diversity. Now research indicates that a significant fraction of the population could reduce their risk of heart disease and cancer by taking statins.

Discovered genes are not novel inventions, and increasingly, with computerized models of proteomic expression, their function is obvious. A democratic transhumanist approach would support the international campaign to ban patents on discovered genes. All discovered genes, plant, animal and human, should be considered the common property of humanity. Countries should refuse to recognize U.S. and European gene patents, and challenge the Agreement on Trade-Related Aspects of Intellectual Property Rights (TRIPS) before the World Trade Organization.

The European acceptance of gene patents appears especially contradictory since European conventions argue that the human genome should not be modified since it is the "common heritage of mankind." At the same time, under enormous pressure from the biotech industry and the United States, Europeans are granting and enforcing gene patents. If the genome is a "common heritage" then it shouldn't be patented, as the EU Parliamentary Assembly has pointed out.

Newly invented genes are another matter, however. The U.S. Patent and Trademark Office first granted a patent on a created life-form in 1980, when the U.S. Supreme Court allowed the scientist Manas Chakrabarty to patent bacteria he had genetically engineered to eat oil spills. Since then hundreds of thousands of patents have been issued for genetically engineered plants and animals, in the United States and the rest of the world. One of the more famous patented mammals is the "onco-mouse," a strain of mice engineered to be cancer-prone to facilitate cancer research. Now genetically engineered human cells are also being patented such as a line of genetically engineered human stem cells that the biotech firm Geron patented in 2003.

Again the stakes for our transhuman future are enormous, since novel gene sequences will be crucial for human enhancement. Some bioLuddites like Rifkin argue that there should be no "patents on life" at all, since this violates the sanctity of the natural order, commodifying Mother Nature, and so on. Forbidding patents on newly created genes would mean that only nonprofit and government researchers would work in this field, which would slow innovation to a crawl.

Within the citizenship–property schema I outlined earlier it makes perfect sense to allow animals without personhood to be owned, and to allow patents on the creation of novel versions of them. We may set aside large swaths of things as common property for aesthetic or ecological reasons, refusing to allow them to be sold. But those things that aren't citizens or common property can be owned by individuals or firms, and when they are inventions, they should be patentable. The processes invented to manipulate human and animal DNA, such as cloning methods or fertility treatments, should also be patentable. Only beings with personhood are exempt from being property, since we each own ourselves and can't be alienated from ownership of ourselves.

One consequence of the self-ownership of persons is that we each own our own genome, and anybody who wants to use our unique genetic code for medical research or reproduction needs to secure our permission first. If someone clones us without permission that would violate our patent on our own genetic code. If a biotech firm clones our stem cells for therapies, we need to be asked and compensated. Our medical and genetic information should also be private, at least from employers and insurance companies, if not from public health officials with legitimate epidemiological concerns.

A second consequence of self-ownership is that as soon as patented genes, whether natural or created, find their way into a self-aware citizen's body, that person becomes a co-owner of the patent. Any person born with artificial patented genes has to be able to read their own gene sequences, share that information and use it to make children. It would be unacceptable, a form of slavery, if someone had to pay a royalty to have a child with their own genes. The law in the United States and Europe tacitly recognizes this point since a gene has to be isolated outside of the human body before it can be patented. But it is still a human gene sequence, discovered in the human genome, whether it is decoded in a lab or inside the human body. It is only patentable when it is outside the body of a person, as well as novel, nonobvious and useful.

The principle that individuals are co-owners of novel genes that end up in their bodies is also based on the 1987 U.S.PTO ruling that a "claim directed to or including within its scope a human being will not be considered to be patentable subject matter" since the "grant of a limited, but exclusive property right in a human being is prohibited by the Constitution." As mentioned earlier, Jeremy Rifkin and Stuart Newman have tried to force the patent office to specify exactly what a human being is by filing for "humanzee" and "humouse" patents, without success. Under the personhood theory of citizenship I've outlined here, the line the U.S.PTO needs to draw is not between humans and non-humans but between persons and non-persons. The 13th Amendment and human rights law in general need to be reinterpreted to forbid property in persons, not in humans.

It seems appropriate that persons should own their genomes since the genome is an intimate element of our personal identity, and is different from other things we may place inside our bodies or brains. For instance, we can distinguish between our ownership of our base DNA and ownership of the DNA of other tissues transplanted into our body. Even if transplanted tissue becomes as inextricable as our original tissue, the immune system and the law can see the difference. We wouldn't permit a firm to repossess someone's artificial heart, but we wouldn't allow the recipient of an artificial heart to start manufacturing knockoffs just because it is a part of his body. Patents on genetically engineered human stem cells lines or the DNA of embryos used to create replacement organs should be similarly acceptable. But the genetic code of every cell in our body has an especially integral feel.

People made with patented genes might be able to use their genes to reproduce or give them away as part of "fair use," but not be allowed to sell them. The principle we currently use to determine which body parts can be sold and which can't is replaceability. We have fewer qualms with the sale of blood or sperm than we do with the sale of kidneys. When we can easily and cheaply regenerate a kidney—which won't be for some time—what exactly will we be protecting by forbidding sales of kidneys? As transhuman technologies progress, just as critics of "body commodification" like Andrew Kimbrell worry, more parts of our body and brain will be detachable, replaceable and commodifiable, and the law will gradually need to acknowledge that our rights to self-ownership include the rights to sell these replaceable parts. This was the persuasive argument made fifteen years ago by Lori Andrews in her classic essay "My Body, My Property." Andrews wrote: "I am advocating not that people be treated by others as property, but only that they have the autonomy to treat their own parts as property, particularly their regenerative parts. Such an approach is helpful, rather than harmful, to people's wellbeing. It offers potential psychological, physical and economic benefits to individuals and provides a framework for handling evolving issues regarding the control of extracorporeal tissues. It is time to start acknowledging that people's body parts are their personal property."

This sounds to many like capitalism run amok. And I am very sympathetic to the idea that many things, such as public lands, should remain uncommodified, the common property of all. If we own our own genes, and we share 99.5% of our genes with all other human beings, then that shared part of the genome is actually the common property of all human beings. There are also occasions when it makes sense to treat body parts as public property, such as when motorcyclists suffer brain death from head injuries. In Belgium, Austria, Finland, France, Norway and Denmark, "presumed consent" laws allow doctors to transplant organs from the brain-dead unless the deceased has a "don't donate" card. Artificial organs, cloned organs and other therapies will probably make it unnecessary to adopt this policy in the United States, which is fortunate given that few Americans support a "presumed consent" system. But just as we have taxes and powers of eminent domain that allow the community to expropriate private wealth for the good of the community, we may decide to selectively forego absolute self-ownership in the body for the good of humanity, and in acknowledgment of the human genome as shared public property.

VOUCHERS TO THE ENHANCEMENT SUPERMARKET

Using Public Policy to Encourage Enhancement

Transhumanists reject any coercion in genetic decision-making. If you don't want to live longer or be smarter, no problem. But that doesn't mean that a transhuman democracy shouldn't encourage people to take advantage of enhancements for themselves and their children. If the current generation is obliged to enact public health measures to ensure the greatest possible health and abilities in the next generation, and there is no ethical difference between treatment and enhancement, then eventually we will need to adopt policies to encourage people to adopt genetic enhancements. This is the conclusion of the profoundly important collaborative statement of four of America's leading bioethicists—Allen Buchanan, Dan Brock, Norman Daniels and Daniel Wikler—in their 2000 book *From Chance to Choice: Genetics and Justice.* In this landmark argument these four consider the many arguments against enhancement and conclude in favor of the "permissibility of rights-respecting genetic perfectionist policies":

> Through its democratic processes, a liberal society could decide to devote resources to the continual enhancement of desirable human characteristics—to embark on a process of genetic perfectionism—so long as in doing so it did not compromise its commitment to justice and the prevention of serious harm. Such a policy need not infringe on individuals' reproductive freedom, for example, if it only encourages rather than coerces or unduly pressures prospective parents to use enhancement technologies. . . . There is no basis for a blanket prohibition of any public policy initiative designed to promote the use of genetic technology for improvement rather than just for the prevention of disease or disability.

In fact, say the four, society has an affirmative obligation to encourage safe and effective genetic screening, therapies and enhancements.

The use of germinal choice and genetic enhancement to ensure greater health and ability is warranted by three concerns. First, we are obliged by our concern for those citizens whose cancers and dementias we prevent, and whose fullest potentials we enable. Second, we have obligations to reduce the burden and costs of disease and stunted ability on society. Third, we have an obligation to encourage as wide an adoption of enhancements as possible in order to minimize the inequalities that result from both natural genetic endowments and unequal access to genetic enhancements.

Any enhancement that promises to make people so dramatically superior in intelligence, longevity or health that it threatens social justice is an obvious candidate for subsidies and universal provision, not for a ban. As bioethicist Peter Singer concludes in his essay "Shopping at the Genetic Supermarket," since a "free market in genetic enhancement will widen the gap between the top and bottom strata of our society, undermine belief in equality of opportunity, and close the 'safety valve' of upward mobility," then society should subsidize genetic en-hancement services so everyone can afford them. We won't be able to, or want to, subsidize all enhancements, however; some will have to be left to be purchased "out of pocket" in the genetic marketplace. So how should we decide which enhancements should be in our universal health care plans, and which should be in the market?

A first step in answering this question is to dispense with the idea that we are only obliged to help the sick and disadvantaged. We have an ethi-cal responsibility to help one another achieve our highest potentials, not just the current human average. The enhancement of ordinary to extra-ordinary abilities is as much a social good as the correction of a disability or illness. When we educate children we don't devote all our resources to kids performing below average because we also want the brighter kids to achieve as much as they can. We don't devote all our Olympic resources to the para-Olympics because we are also delighted to see how much the most talented can achieve. According to the World Health Organization, the mission of medicin e "is a state of complete physical, mental and so-cial well-being and not merely the absence of disease or infirmity." So at least our first cut should be that enhancements should be included "in the plan" along with therapies.

One method of setting health spending priorities is the Quality-Adjusted Life Year or "QALY," which has been applied, more or less, in the Oregon Medicaid system. The QALY idea proposes that the goal of social spending should be to maximize the amount of healthy, happy life produced for each dollar spent. If one year of healthy, happy life is worth 1.0 and being dead is 0.0, then having different kinds of illnesses and disabilities is a quality of life between 0.0 and 1.0. If a thousand peo-ple live seventy years in perfect health, that community produced 70,000 QALYs. But if they all are burdened by disease and pain, and have their lives shortened, then they create many fewer QALYs.

Next, we introduce specific diseases and their treatments. If a person with untreated lung cancer can expect to live one year in pain (say, 0.5 QALY) and the average lung cancer patient who undergoes treatment lives three years in pain (experiencing about 1.5 QALYs), then the lung cancer treatment buys a patient 1 QALY. The per QALY cost of that treatment can then be compared to a treatment for childhood leukemia, which gives an excellent prognosis of complete recovery versus certain death without treatment. Treatment for childhood leukemia might produce something like 50–70 QALYs for the same cost as the lung cancer treatment. Based on this comparison it is clear that we should prioritize treatment for childhood leukemia over lung cancer treatment. We may still pay for both out of public monies, but if there had to be a choice we would pay to treat the child, and ask the lung cancer victim to pay for their own treatment "out of pocket."

QALYs help set priorities for our subsidies for preventive, acute and end-of-life treatments. Since we spend a majority of our health care resources on caring for people in their last year or two of life, the QALY framework argues strongly for shifting spending to preventive medicine that extends healthy life. Again, that's not to say we won't spend resources to try to cure cancers with poor prognoses or spend nursing resources on the terminally ill. But since every dollar we spend on prenatal care reduces our need for neonatal intensive care by three dollars, then we probably need to spend more on prenatal care.

The QALY method applies equally well to enhancement technologies (Figure 12.1). A genetic or pharmaceutical or nanotech intervention that adds fifty years of additional healthy life for a child or an adult is an enormous addition of QALYs whether it adds those years from ages 20 to 70 or from ages 70 to 120. And such an enhancement would compare pretty favorably to most preventive medicine or medical therapies at whatever cost.

Enhancements that improve abilities, intelligence and our daily quotient of joy redefine the goalpost of "1.0." As the WHO definition of health points out, why should our expected quality of life, 1.0, be just the *average* instead of the *best* that we know is possible? The average person in the United States is overweight, has an IQ of 100 and is miserable half the time. In a transhuman QALY scheme 1.0 would be having, at

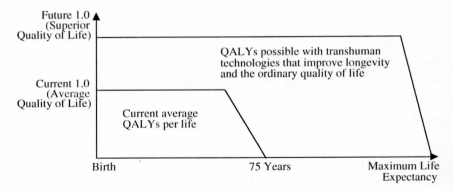

FIGURE 12.1   Calculating QALYs in a Transhuman Health Care System

least, Michael Jordan's musculature, Stephen Hawking's brain and the Dalai Lama's twinkle. Enhancements will raise the bar on all the things that health care workers and patients use to define quality of life:

- ability to perform social roles
- physiological condition, including mobility, exhaustion and pain
- emotional condition, including stability, anxiety, depression and powerlessness
- cognitive condition, including alertness, memory and decision-making ability
- sense of well-being, including pain, interest and lethargy.

BioLuddites will seize on this raising of the 1.0 bar, this quality of life inflation, and argue that it harms the disabled and unenhanced by making their lives seem less rich and joyful by comparison, and by reducing their chances at succeeding in life. This complaint applies equally well, however, to good nutrition or the use of anesthesia: Imagine how people who suffered from scurvy or who endured amputations by biting a bullet felt when they discovered all that could have been avoided. Does teaching people to read do an injustice to the illiterate? Inequality needs to be redressed by ensuring equal access, not by refusing to allow people to live long, able, happy lives because it makes those with short, less able, miserable lives envious.

Another complaint often made of the QALY scheme is that it is impossible to figure out how much each person in a society would value being in each possible state. In particular, the healthy and able-bodied tend to underestimate the quality of life of those who are sick and disabled. We all have an enormous capacity to find happiness in life despite our situations. But if we really believed that, in the absolute scheme of things, it was just as good to be blind, lame or sick then we wouldn't try to reduce disease and disability at all. If QALYs are assessed and priorities set in an open democratic process we can take account of the voices and experiences of the sick and disabled, and still set policies that aim to reduce disease and disability.

Oregon's experience demonstrates that the QALY scheme can be the basis of an open, democratic priority-setting process that includes a full range of voices and values. Oregon developed its priority weights on disease-treatment pairs through many years of public meetings, surveys and legislative debate. Other values, such as the egalitarian preference for the needs of the disabled and poor, were addressed in that debate and included to adjust the equations.

Without egalitarianism as part of the equation the QALY scheme could be interpreted to argue that giving someone with a genius 140 IQ ten more points is as valuable as giving someone with a subpar 90 IQ ten more points. But intuitively we understand that giving greater priority to treatments for the poorest, sickest and most disabled will generally produce more happiness for those individuals, and more good for society as a whole. Even with a little QALY affirmative action for the treatments that benefit the poor, sick and disabled, treatments to enhance ordinary intelligence, happiness or strength should still be covered under the scheme if they are cheap and beneficial enough. Otherwise, those enhancements could be purchased out of pocket in the marketplace.

A QALY schema would also help adjudicate which enhancements should be actively discouraged. For instance, the transhumanist writer Nick Bostrom distinguishes between intrinsically rewarding enhancements, which benefit the individual no matter how many people receive them, and "positional enhancements," which only benefit an individual if they are not widely shared. The intrinsically rewarding enhancements should be encouraged and subsidized, he argues, while positional enhancements should at least be discouraged if not banned. Greater intelligence or longevity are intrinsically rewarding enhancements, so that both the people who receive those gifts and the society that allows people to choose them are better off. Even if everybody else gets smarter, it is still better to be smarter than less smart, and a smarter society is better off overall. But if parents widely used genetic modification to make their children taller, which has no *intrinsic* value but is only valuable if your child is taller than their peers, they would be taking some level of risk for no overall social benefit. Subsidizing enhancement that brought the shortest 10% of kids up to average height would produce some QALYs, but subsidizing height enhancement for a majority of children would produce no QALYs and might involve health risks that would rob society of QALYs. A policy permitting height enhancement for the shortest 10–20% would still, however, be far more generous than the U.S. Food and Drug Administration restriction that human growth hormone only be prescribable for the shortest 1% of children.

Selecting a child's sex, sexual preference or skin color does not produce QALYs. These germinal choices should be permitted in a society that values procreative liberty, but need not be subsidized in the service of procreative equality. Choices that actually *reduce* QALYs, such as cutting off your own healthy leg, or choosing a deaf embryo over a hearing embryo, or refusing simple, safe treatments that increase a child's quality of life, should be strongly discouraged by society. We may value individual liberty and parental prerogatives so highly that we stop short of banning these choices. But, as with smoking and other social vices, we can ratchet up the pressure through public education and professional self-regulation, and by obliging people to defend their choices as sane and consistent with strongly held beliefs such as religious faith. Society need not remain neutral toward bad choices just because it permits us the freedom to make them. Figure 12.2 presents the range of regulatory options that might be applied in a transhuman democracy.

> **Mandatory, no exceptions,** e.g., emissions controls on cars, and quarantine of dangerous infections
>
> **Required, but with religious exemptions,** e.g., the military draft, vaccinations
>
> **Publicly funded and encouraged, but not obligatory,** e.g., public school
>
> **Encouraged, but unsubsidized,** e.g., exercise
>
> **Publicly funded, but not encouraged,** e.g., birth control or abortion in some states
>
> **Available unsubsidized on the market,** e.g., over-the-counter drugs
>
> **Available, but controlled by prescription,** e.g., Prozac, Ritalin, etc.
>
> **Discouraged, by taxation and education, but not banned,** e.g., smoking
>
> **Banned,** e.g., illegal drugs, some weapons, dangerous toxins

FIGURE 12.1   Options for Regulating Technologies in a Liberal Democracy

GETTING A PLAN

So the QALY scheme is just the starting point for a public, democratic discussion about which technologies should be "in the plan" and universally available. QALYs give us a framework for talking about the consequences for liberty and equality of different distributions of technology. Cynics will respond that spending priorities in the real world are determined by the best-financed political action committees and the most expensive lobbyists. The United States doesn't even have a universal health care system, despite consistent public support over the last century, because of the opposition of doctors and big business.

Which is all too sadly true. But, as Jane Addams said, the solution to the problems of democracy is always more democracy. In every other industrialized country, people have to struggle within their parliaments for generous funding and rational priorities for health care. In the United States we have to wage a two-pronged struggle, to create an equitable, universal and cost-effective health care system, and to ensure that it generously provides all beneficial technologies including enhancements.

One of the principal complaints about a universal health care system, and attractions of the libertarian free-market vision, is that markets are better at providing individual choices and satisfying individual preferences than governments. But it is possible to imagine a universal, egalitarian health care system that makes enhancements widely available while accommodating radically different values, and radically different calculations of QALY. The answer is to permit consumers to purchase health care from competing health plans with a voucher. This was actually the system proposed by President Clinton in 1994, and bioethicist Ezekiel Emanuel gives the idea of a universal voucher system with competing, radically different health plans a very interesting defense in his book *The Ends of Human Life*. Under such a system Catholic hospitals and health care workers could form their own Catholic health care network, and enroll consumers who don't want contraception, abortion, euthanasia or embryonic stem cells. The money they saved could then be spent on extensive end-of-life care. Other plans could choose to provide access to germinal choice, stem cells, artificial organs, cybernetic enhancements and cryonic preservation for families who valued them, by, for instance, reducing care for the permanently unconscious. So long as every citizen is assured a basic set of health benefits, and all those who want enhancements for themselves or their kids have the choice to use their health care voucher to buy into a plan that provides enhancements, the system would provide both equity and choice.

But even if the rich do get more enhancements in the short term, it's probably still good for the rest of us in the long term. The first personal computers were expensive and difficult to operate. Eventually usability improved, prices dropped and personal computers became accessible to people in the developing world. If the wealthy stay on the bleeding edge of life extension treatments, nano-implants and cryo-suspension, the result will be cheaper, higher-quality technology. The poor bore the brunt of nineteenth-century medical experimentation. Perhaps the rich should take a turn as guinea pigs this time.

13

Defending the Future

The bioLuddites are unhappy with the idea of transhuman technologies being regulated under existing agencies, such as the U.S. Food and Drug Administration, which are mandated only to ensure safety and efficacy. Instead they want new agencies empowered to ban technologies on the grounds of vague long-term risks such as the future conflict between humans and posthumans. Consequently transhumanists should promote the use of the existing agencies to regulate transhuman technologies in order to protect more liberal access.

Nonetheless, we do need to take some long-term risks seriously, especially the problem of human–posthuman solidarity. In this chapter I review several ways of working to minimize the threat of a human–posthuman schism. One is to make societies highly tolerant of diversity so that they might be better prepared to accommodate transhuman diversity. Another is to build global institutions that protect civil rights and keep the peace. A third is to ensure that all beings allowed powers significantly greater than those of unenhanced human beings are endowed with empathy for humanity and a moral code. I discuss the relationship of empathy and moral reasoning to citizenship, and the challenge of respecting cognitive liberty in a future in which our emotions, ethics and beliefs can be directly examined.

REGULATING FOR SAFETY NOT HUBRIS

The ecological movement has grown increasingly anxious about the health and environmental risks of nanomaterials, and for legitimate reasons. Most of the 87,000 industrial chemicals in use have never been fully tested for their toxicity. Evidence has mounted that a class of these industrial chemicals mimic the hormone estrogen, disrupting the endocrine system and causing birth defects. Endocrine disruption during embryonic development is implicated in falling sperm counts, rising rates of male infertility, structural deformities of the genitals, testicular cancer and impairment of neurological and immune functions, both in humans and in animals.

Under the Clinton administration the Environmental Protection Agency (EPA) began a program to complete the testing of the 4,000 or so chemicals produced in quantities of a million pounds or more each year. Less than half of these chemicals had been tested for the principal hazards—their acute and chronic toxicity, their effects on reproduction and development, whether they cause cell mutations and cancer, and their effect on the environment. Only 21 of the 830 companies making these chemicals in the United States had conducted all these tests on all their chemicals. The Clinton administration ordered the EPA to work with chemical manufacturers and the environmental agencies around the world to speed up the testing. Fortunately, like human enhancement, toxicological and epidemiological research is also benefiting from the rapid advance of the molecular and computational sciences. The EPA's endocrine disruption research program is now using "computational toxicology" to prioritize which chemicals need to be tested for endocrine effects. "Oncomice" are making it easier to test for carcinogens.

But even complete toxicity testing of each chemical in isolation won't really satisfy environmental and health questions since the chemicals behave differently when they mix in our bodies. In 2003, when the Centers for Disease Control and Prevention randomly tested the blood and urine of 2,500 Americans, they found 89 different industrial chemicals, including pesticides, herbicides, pest repellents and disinfectants. The concentration in most people's blood was below the toxic level for each chemical, but the cumulative effects are unknowable.

So although we can never be certain of the safety of any new material, it is clear that we need to expand the testing, control and cleanup of industrial chemicals nationally and internationally. Unfortunately the chemical industry suppresses and falsifies information about the toxicity of chemicals, and pays off (mostly Republican) legislators with hundreds of millions of dollars of contributions. Corporate-deformed democracy allows industry to "externalize" the environmental and health costs of production. The Bush administration in particular has devoted itself to an aggressive erosion of the EPA's mandate and administrative easing of environmental regulations for farmers, oil drillers, miners, industry and the military. The latest Bush budget calls for reductions in funding of research on endocrine disruptors, pesticides and toxics. The problem is not the abundance of technology but the deficit of democracy.

New nanomaterials need to be subjected to the same strict toxicological scrutiny as these other chemicals and materials. But there is nothing distinctly different about nanotechnology compared to other industrial materials that justifies the demand by some extreme environmental groups for a global "precautionary principle" moratorium on nanotechnology. In fact a report released by Greenpeace in August 2003, "Future Technologies, Today's Choices," looked at the risks and benefits of nanotechnology and concluded that while industry and government should commit to testing nanomaterial safety, no new laws or regulatory agencies are required to govern nanotechnology. In the United States, for instance, the EPA has the authority under the Toxic Substances Control Act to require that nanomaterials be tested, to gather information about their safety and use and to issue regulations for their control. Concerns about more distant risks, such as gray goo, are being addressed by the Foresight Institute and Center for Responsible Nanotechnology with designs for "limited molecular nanomanufacturing" that can't get out of control.

In general, the regulatory agencies required to ensure the safety of transhuman technology are already in place, and we just need to make sure they are given the funding, authority and independence from corporate influence to do their job. But the bioLuddites don't want to give the EPA or the Food and Drug Administration the mandate to regulate the safety of transhuman technologies. They want agencies like the United Kingdom's Human Fertilisation and Embryology Authority (HFEA), which is empowered to forbid things because of vague social anxieties.

For instance, in 2003 the HFEA ruled that a family could not test the DNA of their invitro fertilized embryos to decide which one to implant. The parents wanted to make sure that the baby would be a tissue match for his or her 3-year-old brother, Charlie, who had a rare and lethal blood disorder. With stem cells from the new baby's umbilical cord they might be able to save Charlie's life. But the HFEA turned them down, arguing that would be the first step toward "designer babies." The HFEA pondered "the psychological burden that may be placed on a child who was an 'engineered' match." But they didn't consider the special bond that might exist between siblings when one had saved the other's life. So the Whitakers had the procedure done in the United States, which does not yet have an HFEA. Hundreds of parents are now having savior baby testing done around the United States.

Although we are more liberal than Britain and some parts of Europe, the U.S. government has been regulating the safety of human genetic engineering. It first began regulating genetic engineering back in 1974 when the National Institutes of Health established the Recombinant DNA Advisory Committee (RAC). The RAC ensured that genetically engineered organisms were handled safely and weren't released into the environment. It developed guidelines for human gene therapy trials in 1983, well before any trials were carried out. All gene therapy protocols funded by the NIH are reviewed by RAC, and some privately funded gene therapy research is voluntarily submitted to RAC review. The NIH examines both the safety of the clinical trials as well as other ethical issues. In 1997, the NIH convened its first conference on genetic enhancements, and the RAC has announced that it will not accept proposals for germline genetic modifications. So the RAC already fulfills the demand of many bioLuddites that there be public debate and due caution before we permit germline therapy or enhancements.

But the RAC has only statutory power to review publicly funded research. The Food and Drug Administration has the wider mandate under the Federal Food, Drug and Cosmetics Act and Public Health Service Act to regulate pharmaceuticals and medical devices. Before a new drug or medical therapy can be used on humans, whether developed with private or public funds, it must pass through the clinical testing regime of the FDA. The FDA announced its intention to regulate human gene therapy in 1984, and its Center for Biologics Evaluation and Research (CBER) oversees and regulates genetic research. Researchers who want to test genetic therapies in humans have to prove to the FDA that animal studies demonstrate an adequate level of safety for Phase I, II and III trials. After the death of Jesse Gelsinger in 1999 in a University of Pennsylvania gene therapy experiment, the FDA closed down trials at institutions found not to be in compliance with FDA safety rules. The FDA has also asserted that human reproductive cloning trials must go through their approval process, and that they have no plans to approve any reproductive cloning since it is still too unsafe.

The bioLuddites assert that the FDA's mandate is not broad enough to regulate the new genetic and reproductive therapies. But the FDA has repeatedly insisted that their mandate is broad enough and most legal observers agree. Attorney Christine Willgoos concluded in a 2001 review of FDA regulation of cloning and germline therapy published in the *American Journal of Law and Medicine* that "the FDA already has appropriate mechanisms in place to respond to the concerns surrounding cloning and germline gene therapy technology." As the President's Council on Bioethics pointed out in 2004, the FDA's mandate may need to be broadened a bit to include invitro fertilization, pre-implantation genetic diagnosis and other fertility techniques, which, unlike fertility drugs, aren't required to go through FDA approval. But this function does not require a new agency. As galling as it is to the libertarian transhumanists and ban-minded bioLuddites alike, the FDA, EPA and the equivalent regulatory agencies around the world are our best hope for a liberal public policy toward human enhancement technologies.

ENSURING SOLIDARITY AND OUR RIGHT TO MISANTHROPY

We might create a group of people much smarter than us, that might want to kill us. Or we might want to kill them. If we can't go 100 years without a genocide, then we have no business altering the species.

—GEORGE ANNAS,
quoted in "Will Genetic Engineering Kill Us?"
Wired, April 16, 2003

[Science] is man's gradual conquest, first of space and time, then of matter as such, then of his own body and those of other living beings, and finally the subjugation of the dark and evil elements in his own soul.

—J.B.S. HALDANE,
"Daedalus, or Science and the Future"

Siding with the X-Men

In his 2002 piece "The Transhumanists" in the *National Review Online*, Wesley J. Smith argues that "transhumanism envisions a stratified society presided over by genetically improved 'post-human' elites. Obviously, in such a society, ordinary humans wouldn't be regarded as the equals of those produced through genetic manipulation."

Hopefully it is clear by now that the subjugation of humanity is not on the transhumanist agenda. As I propose in the previous chapter, it is easy to imagine a tiering of rights in a transhuman society that does not allow posthumans any more rights than humans, and social policies that avoid polarization by making enhancements widely available. Occasionally transhumanists have confirmed the worst fears of critics like Smith by expressing indifference to the future of humanity. For instance, the computer scientist Hans Moravec has written that he doesn't care that robots will replace humans since evolutionary succession is the natural order of things. Some extropians have expressed a similarly Nietzschean disregard for the fate of the average human. Extropian misanthropy was memorably satirized by left-wing transhumanist Ken Macleod in his novel *Cassini Division,* in which extropians turn Jupiter into a supercomputer, upload themselves and launch attacks on a socialist solar civilization.

The X-Men mythos provides another popular science fiction metaphor for a potential human–posthuman schism. In the X-Men universe the conflict is between the posthuman supremacist Magneto, who believes posthumans should rule over humans, and Professor Xavier, who believes humans and posthumans should live in equality and solidarity. In the first issue of the X-Men comic in 1963, Jean Grey asks Professor Xavier "Just what exactly is our real mission?" Xavier answers: "Jean, there are many mutants walking the earth and more are born each year. . . . Some hate the human race and wish to destroy it! Some feel mutants should be the real rulers of earth! It is our job to protect mankind from those . . . real mutants!" In the next panel the leader of the evil mutants, Magneto, rants "[I shall] make homo sapiens bow to the homo superior."

The potential development of a posthuman elite hostile to humanity was addressed in the WTA's 1998 Transhumanist Frequently Asked Questions 1.0 ("FAQ1"). FAQ1 says that ensuring the mutual tolerance, social solidarity and political equality of humans and non-humans is a central mission of transhumanism. But it also acknowledges that posthumanity may become so radically different that coexistence with humans would be impossible, and that then, ideally, posthumans should be able to separate or migrate to space. But "if the posthumans are not bound by human-friendly laws and they don't have a moral code that says that it would be wrong, they might then decide to take actions that would entail the extinction of the human species."

FAQ1 suggests three measures to ensure a "Xavier" outcome. First, we are urged to promote tolerance of diversity and compassionate solidarity. That will build the kind of society that will be ready for transhuman diversity, and inculcate appropriate values in those who eventually get enhanced. Second, the world needs to "build stable democratic traditions and constitutions, ideally expanding the rule of law to the international plane as well as the national." Those global institutions will protect human and posthuman rights, enact policies—such as technology transfer—to reduce conflict, and intervene when conflicts break out.

Third, FAQ1 urges that "values of tolerance and respect for human well-being are incorporated as core elements of the programming" of machine minds. The authors are explicitly advocating the work here of people like computer scientist Eliezer Yudkowsky and his Singularity Institute for Artificial Intelligence, to design pro-human "friendliness" into AI. But why propose that machine minds be shackled by pro-human sentiments and morality, while posthumans are only exhorted to be nice and punished if they act badly? Wouldn't it be more consistent, and safer, to encourage posthumans to adopt the same moral and emotional safeguards that the FAQ1 proposes for the machines? Is it possible to imagine a "liberty-respecting" policy that discourages misanthropy among posthumans?

EMPATHY AND MORAL REASONING

We may be able . . . to control our passions by some more direct method than fasting and flagellation . . . to deal with perverted instincts by physiology rather than prison.

—J.B.S. HALDANE,
"Daedalus, or Science and the Future"

So who is to tell us that being human and having dignity means sticking with a set of emotional responses that are the accidental byproduct of our evolutionary history? . . . Why don't we simply accept our destiny as creatures who modify themselves?

—FRANCIS FUKUYAMA,
Our Posthuman Future

The problem with designing friendliness or empathy into machine minds is that we don't yet understand friendliness and empathy in the organic human mind. We are still debating whether altruism, empathy and sociability are mostly sociobiological or mostly socialization. The sociobiological view of empathy is central, for instance, to Fukuyama's complaints about human enhancement. Fukuyama believes that if we allow people to change any of the biological features that make up the "Factor X" of humanness we will become too different to feel part of the same racial community. But if social solidarity is rooted in the sociobiological instinct to help the reproductive success of people who share our genes, then why would we feel more solidarity with someone of a different race on the other side of the planet than with our cousin who got some life extension and intelligence tweaks? Basing solidarity on biological similarity is as much a formula for conflict within the human race as between the human race and posthumans.

In fact, Fukuyama specifically defends instincts to dominance, aggression and tribalism as part of human nature and suggests they have adaptive importance: "What if you decide you don't like aggressive children, since aggression is the root of violence and school shootings, and so forth? So, either through drugs or through some sort of genetic intervention, you discover the source of this kind of aggression. Then you no longer have either innovation or people just standing up for principle, because it turns out that in the human psyche it all comes from the same source." Fukuyama is specifically concerned that socialist experiments, supposedly defeated by innate selfishness and clannishness, might succeed with the help of enhancement technologies that removed those propensities.

Even if the root of social solidarity is a sociobiological impulse to identify with similar creatures, that doesn't really preclude solidarity with chimps and posthumans, and doesn't naturally suggest the brotherhood of man. So why shouldn't the impulse also be expanded beyond human-racism to solidarity with "the well-being of all sentience" as Buddhism, transhumanism and the animal rights movement advocate? If the impulse is that malleable, perhaps we can just set aside altogether the idea that our morals spring from our genes and focus on the empathy we feel for other creatures.

Empathy is the ability to imagine that someone else feels the kind of emotions that we do, to place ourselves in their shoes. Since we want our own lives to continue, we can sympathize with others who also don't want to be killed. Empathy is therefore intrinsic to the capacities that make us persons in the first place: self-awareness, self-concepts, desires and abstract reasoning. Empathy is so closely associated with the capacities necessary for legal personhood that theorists like Joseph Fletcher, Daniel Dennett, Tristram Engelhardt and Joel Feinberg have included empathy and sensitivity to moral censure as necessary elements of personhood.

Many lines of research are converging to suggest the centrality of empathy to personhood. Howard Gardner has argued that "emotional intelligence," including an understanding of one's own and other's feelings, is separate from and equal in importance to the abilities tested on IQ tests. Researchers like neurologist Antonio Damasio have shown that our emotions and our response to the emotions of others are much more intimately involved in our reasoning than had been previously imagined. Neuroscientists Arthur Craig and John Allman have identified "spindle cells" and the "insulae" as brain structures that process our sense of self, our internal monitoring of our own existence, as well as our feelings of outrage at lying and injustice and our feelings of love and shame. These structures are activated by both rational calculations and moral decision-making. We share these structures most closely with the great apes. Human babies only develop spindle cells in their fourth month, and they gradually grow after that.

It makes sense then for a society to agree that all mature persons should have the capacity for empathy. Persons without empathy are disabled citizens who need our help to be returned to the full cognitive capacities required to be fully self-governing and responsible citizens. Disruptions in one of the brain processes necessary for recognizing other beings' emotions, associating them with one's own and feeling some kind of empathic response lead to asocial or anti-social behavior. For instance, autism is a brain disorder one of whose symptoms is an inability to understand or interpret other peoples' feelings. Although the autistic can have wonderful compensatory savant abilities, just as with blindness and other disabilities, society has an obligation to ensure that as few people as possible suffer from this disability, and that we try to find a cure. Although the autistic are generally too impaired to be harmful to themselves or others, other kinds of brain damage and genetic predispositions have been linked to aggression and sociopathic behavior, and there is an even stronger case for their prevention and cure.

Short of severe mental disability, research shows that everyone's personality can be described as a mix of five basic characteristics, all of which are substantially inherited and stable across one's life: Openness to Experience, Conscientiousness, Extraversion, Agreeableness and Neuroticism. One of the "five-factor" researchers, Kerry Jang at the University of British Columbia, has found that agreeableness or sociability, and associated beliefs and attitudes, are especially strongly influenced by genes. People who score high on sociability are more compassionate, trusting and helpful, whereas people low in sociability are uncooperative, unsympathetic and easily irritated. Jang also finds these personality traits related to political attitudes. For instance, people who inherit high sociability were more likely to favor gender equality and open-door policies for immigrants.

As philosopher Mark Walker suggests in his essay "Genetic Virtue," society could reap tremendous benefits by identifying the genes and neurochemicals necessary for empathy and cooperation, encouraging noncoercive screening and therapy to ensure that all citizens have them, and giving incentives for people to select for them in children and amplify them in themselves. If we find that our innate drives are far more selfish than altruistic, more tribal than universal, more hierarchical than egalitarian, we can encourage one another to use brain technologies to temper these impulses. Far from making us less human, moral self-improvement may be our most radical assertion of human reason and achievement. Referring to his theory that culture is coded in "memes," Richard Dawkins wrote in *The Selfish Gene* that "We have the power to defy the selfish genes of our birth and, if necessary, the selfish memes of our indoctrination. We can even discuss ways of deliberately cultivating and nurturing pure, disinterested altruism—something that has no place in nature, something that has never existed before in the whole history of the world. We are built as gene machines and cultured as meme machines, but we have the power to turn against our creators. We, alone on earth, can rebel against the tyranny of the selfish replicators."

Conversely, transhuman technologies—genetic, pharmaceutical or cybernetic—will also make it possible to eradicate or suppress impulses to empathy and solidarity. Magneto-ites or Ayn Randians could "liberate" their selfishness from empathy. Although we want to create a society that is as open and tolerant of body modification, germinal choices and brain self-management as possible, there is no reason for even the most liberal society to remain neutral toward whether the next generation has more or less capacity for aggression and selfishness, or empathy and cooperation. We can respect liberty and still encourage people to care about the greater good for the greater number.

Some Left bioLuddites point to the history of eugenics to suggest that efforts to modify the biological bases of anti-social behavior will be directed at the poor, ignoring the crimes of the rich. This wasn't true of all eugenicists, however. The birth control campaigner Margaret Sanger, although a sometime supporter of the eugenics movement, rejected the idea that criminality or race should be connected to eugenic improvement. Instead she thought that the genes that encourage war-making, such as selfishness, would be a better target: "Felonies may be great events, locally, but they do not induce catastrophes. The proclivities of the war-makers are infinitely more dangerous than those of the aberrant beings whom from time to time the law may dub as criminal. Consistent and portentous selfishness, combined with dullness of imagination, is probably just as transmissible as want of self-control."

Nonetheless, the technologies of brain control will probably first be applied to the rehabilitation and control of criminals rather than to the power elite, and we will have to carefully balance liberty against the public good. Yet the civil liberties issues raised by the use of empathy boosters, aggression suppressants and morality chips in criminal rehabilitation are already with us. Many psychiatric disorders cause people to commit crimes they wouldn't have when medicated, and we are still divided over the culpability of the mentally ill for those crimes. Since testosterone is the principal driver of libido and aggression, nine U.S. states mandate that repeat rapists submit to injections that suppress testosterone, or have their testicles removed. Other states make chemical or surgical castration optional, but link it to earlier parole. Studies find that chemical or surgical castration dramatically reduces the return to sexual violence. A study of German rapists who volunteered for surgical castration found that only 3% repeated their crimes compared with 46% for noncastrated offenders. Libertarians and the American Civil Liberties Union have condemned mandatory chemical castration, but the option has been enthusiastically embraced by many offenders who are happier when freed from their obsessive, violent impulses.

It will take much passionate debate to define the appropriate ways to encourage empathy and cooperation while avoiding liberty-eroding coercion and campaigns against thought-crime. But democratic societies can and should encourage sociability and empathy for all sentient beings. Subsidized prenatal care could provide genetic screening for gene markers for autism, psychopathy, aggression and impairments in the cognitive capacities necessary for empathy. The health care system can then make available genetic, pharmaceutical and nanomedical therapies to correct those deficits. Just as we ask people to submit to driver's tests and ask potential employees to offer proof of citizenship, people could be asked for proof that they have fully functional capacities for empathy. Conversely, laws may be required to stop businesses and the military from selecting for people with *lower* capacities for empathy, or providing technologies that facilitate the robotic pursuit of orders or profits.

ELECTRONIC SUPEREGOS

Some sociobiologists believe that even our innate understanding of social justice may have a genetic basis. Swiss researchers Ernst Fehr and Simon Gächter have been exploring the apparently innate drive we have to punish people who cheat society, which they call "altruistic punishment." In experiment after experiment people will go out of their way to punish people who break the rules of games, hoard or otherwise flaunt cooperative expectations. According to Fehr, "The emotional satisfaction of dispensing justice seems to spur them on."

Punishment of "free riders," people who benefit from social cooperation without cooperating themselves, has long been recognized as an essential feature of social organization. But Fehr and Gächter's work suggests that the drive to punish free-riders may be innate, which makes perfect sense in terms of the "group selectionist" account of genes. If there is group selection, altruistic genes good for the group's chances for survival would be selected for even if they reduce the individual's likelihood of reproducing.

Empathy, a pro-social orientation or the impulse to punish free-riders, by itself is the emotional driver of moral behavior. In order to behave morally we also need moral reasoning, an understanding of the rules of conduct and the consequences of our actions. Is it OK for someone to steal to feed their kids? Empathy for the shopkeeper might say no, empathy for the parent and kids might say yes, and empathy for the society as a whole say no again. We need some model of decision-making to decide which way our empathy should cut.

In the 1970s, Harvard psychologist Lawrence Kohlberg discovered that children and adults develop moral reasoning through six stages. Each stage is more sophisticated than the one before, and someone at a latter stage incorporates the insights from the previous stages. The first stage is pure selfishness, anything that the child wants is what is right. That gets generalized to a Stage Two ethics of whatever anyone wants is what is right. You scratch my back, I scratch yours, we're all happy. Stage

Three defines morality as good intentions, and Stage Four defines morality as obeying laws and protecting the social order. People at Stage Five reason that we all have obligations to follow the rules of a democratically based social contract that protects individual rights and interests. People at Stage Six are committed to a democratic social contract but also to universal ethical principles that can oblige acts of conscience such as civil disobedience.

Kohlberg was accused by some scholars of political bias since his final stages are basically liberal democratic values. But he found that children and adults from around the world worked through the first five of these stages in the same way as in the United States. Advanced moral reasoning was correlated with age, education and the complexity of the social structure, and probably also exposure to liberal democratic thought. Unfortunately Kohlberg found that very few adults anywhere reached Stage Six, and few of those used universal ethical principles consistently.

I carry around a palm pilot that compares what I eat to my dietary goals for carbohydrates and calories. In effect, it whispers warnings in my ear when I eat something I'll regret. Once we have intimately wedded cybernetics to our brains, and have onboard expert systems watching and advising our behavior, we will undoubtedly also develop cybernetic superegos with settings for Kohlberg's Stage Six, Islamic Sharia or Ayn Randian selfishness. We may give children this mental equipment, and the ability to modify it as they mature, just as we currently provide them with moral and civic education. We might then all be able to consistently reason with the clarity of philosophers and the selfless compassion of Gandhi or Martin Luther King. "Contemporary Man does not foresee," said George Bernard Shaw in *Man and Superman,* that "the real Superman will snap his superfingers at all Man's present trumpery ideals of right, duty, honor, justice, religion, even decency, and accept moral obligations beyond present human endurance."

In that world, we will be very interested in whether our spouses, coworkers, leaders and police are running moral reasoning support software, if so what kind, and if they also have a healthy sense of empathy or pro-socialness to encourage them to do what they know to be right. Again, it would be a violation of cognitive liberty for society to *require* adults to adopt moral codes, listen to an internal psychic nag or publicly disclose their empathic and moral orientation. But it is acceptable in a free society to make free choices consequential. If you are a sexual predator and you don't want to be castrated then you have to stay in prison. If you want security clearance or a job working with toddlers or flying jets you may have to submit to a background check or even a lie detector test. We require judges, politicians and physicians to take oaths before they assume office because their powers are so easy to violate and we want the faint protection offered by a solemn vow that they will not abuse them. In the transhuman future there may be many occupations in which it is not only a requirement of the job that one have the necessary skills, but also that one have internalized the necessary pro-social feelings and moral reasoning software.

DRIVER'S LICENSES FOR SUPERPOWERS

In such a society we might strike a bargain with humans, animals or computers who want superhuman abilities: Give us some evidence that we can trust you not to abuse those powers. As Spider-Man said on becoming a posthuman, "With great power comes great responsibility." Society could forbid the most powerful enhancements to those who refuse to be screened for basic empathy and morality, just like we do a background check for a handgun purchase. The manufacturers of cognitive enhancement software could be obliged to include empathy and moral decision-making supports as a feature just as we require warnings and child-proof caps on medicine and air bags in cars.

But it is not just in society's interest that we steer enhancements toward empathy and cooperation. It is also an ethical obligation on each of us to enhance ourselves, to become better people and use our powers to do good. Transhumanist philosopher Mark Walker argues in his "Neo-Irenaean" thesis "Becoming Gods" that the goal of human enhancement should be moral as well as mental and corporeal perfection. Walker argues that we are obliged to cultivate our virtues and do more good in the world by adopting specific enhancements such as boosters for our empathy and intelligence. The use of transhuman technology for simple distraction—such as to live in a solipsistic virtual reality or to spend more time perfecting our golf game, or to eliminate our compassion, courage or curiosity—would be unethical even though it should be legal.

On the other hand, policies to encourage empathy and morality should not be so invasive that they inhibit diversity and creativity, or threaten our privacy and freedom of thought. You should be able to get enhanced and still have extreme ideas, have a hankering for parrot meat, or despise whole groups of people.

The lines will be difficult to draw, but just because they will be difficult does not mean that we should either ban powerful enhancements or have a completely laissez-faire approach. The best outcome for society, and probably the only possible path, lies in muddling through to keep the growing diversity of cyborg citizens free, equal and united.

A Sexy, High-Tech Vision for a Radically Democratic Future

What more can be said of any condition of human affairs, than that it brings human beings themselves nearer to the best thing they can be?

—JOHN STUART MILL, *On Liberty*

Whether it amounts to a Singularity or not, the coming decades will turn our world upside down and our expectations inside out. Radical times call for radical solutions. What should the strategic goals of democratic transhumanists be in the coming decades?

BUILD THE TRANSHUMANIST MOVEMENT

The bioLuddites and human-racists are already in place. They control the bully pulpit of the President's Council on Bioethics, they have their mediagenic talking heads, their think tanks, foundation grants, journals and lobbyists. They are organizing conferences for devout Christians, radical Greens and suburban soccer moms. They have filled the media with alarmist rhetoric about *Frankenstein, The Boys from Brazil* and *Brave New World*.

There is no comparable, serious voice, recognized by politicians, academe and the media, articulating the transhumanist case. The World Transhumanist Association and Extropy Institute combined could barely cover the phone bill of any one of the shrill sirens of doom. Will human enhancement only be championed by glib corporate press releases and timid researchers unwilling to project the consequences of their discoveries even five years down the road? Will the only people debating realistic solutions to the problems of the twenty-first century be those who want to keep the twenty-first century from happening at all?

We need transhumanist think tanks, journals, conferences and lobbyists. We need transhumanists meeting the bioLuddites toe-to-toe in the public square, defending the rights of persons to use reason to control their own affairs. We need transhumanist clubs and study groups on the campuses, and in every city in every country, educating the public about the threats and promises to come. We need a movement fighting for a positive future, and not just fighting the future.

BUILD THE NEXT LEFT

There will be a Next Left, and it will be global. We can see its embryonic beginnings in the anti-globalization movement knitting together disparate causes around the world. The Next Left will inevitably be connected to, if also often in conflict with, the existing social democratic and labor parties. Although many of the progressive reform movements are currently hostile to technology, many will make the same shift toward biopolitics and techno-empowerment that the nineteenth-century workers movement made toward industry, from Luddite machine-wrecking to empowerment *through* technology.

The global social justice group the World Social Forum (WSF) says in their statement of principles that they are a movement "committed to building a society centred on the human person." Radical democratic movements of the twenty-first century, such as those involved with the WSF, have learned from the twentieth-century authoritarianisms to make radical individual liberty a core principle, and to derive the need for equality and public provision from the need to maximize individual potential. Having adopted this vision of radical individual empowerment, these movements will soon see the importance of universal access to human enhancement technologies. The Next Left will wield the same values—liberty, equality and solidarity—as its democratic predecessors, but this time it will seek to create a democratic community that includes *more* than just all humanity.

RADICALIZE "HUMAN RIGHTS"

Defend the rights of all beings oppressed because of their bodies

Democratic transhumanists need to build solidarity with those oppressed because of the bodies and minds they possess. A world tolerant of transhuman diversity is best assured by expanding the bounds of tolerance and equality to include the full diversity of already existing beings, sexual, cultural and racial. Racism and discrimination in all forms must be opposed. The physically disabled should have access to the social and technological assistance they need to be equal citizens. Gender must not determine rights, so civil marriage must be open to gays and lesbians. People should be allowed to use technology to sculpt themselves to fit their personal visions, whether they fit the binary gender system or not. Democratic transhumanists should join the campaigns to extend rights to great apes, dolphins and whales in order to defeat human-racism and open rights for all intelligent persons.

Guarantee the right of all persons to control their own bodies and minds

We need not only a broader concept of the citizen, the bearers of rights, but also a more radical understanding of the rights those citizens can claim. Self-ownership should include the right of sane adults to change and enhance their bodies and brains, to own their own genes, to take recreational drugs and to control their own deaths. Procreative liberty, an extension of the right to control our body and life, should include the right to use germinal choice technologies to ensure the best possible life for our children. Strong democratic government is required not only to protect these rights, but to ensure that the technologies are tested for safety, and that consumers understand their risks and benefits. We need to ensure that all citizens have access to these options, not just the affluent.

DEMOCRATIZE TECHNOLOGICAL INNOVATION

Support science education and public funding of research into
transhuman technologies

We need expanded public financing of higher education and scientific research throughout the world. Higher education, and education in the sciences in particular, helps spread rational, secular values, empowers citizens to make informed choices about technology policy and increases the pool of human talent available to apply to improvements in human welfare. American children in particular are woefully behind the rest of the industrialized world in math, science and engineering. More American students get degrees in "parks and recreation" than electrical engineering.

The National Nanotechnology Initiative (NNI) and its NBIC program are wonderful examples of how basic government research funding can kickstart the development of transhuman technology. The NBIC report recommends a major initiative, like the Human Genome Program and NNI, to understand and enhance the working of the brain with the new molecular and information technologies. Dr. Aubrey de Grey has argued that we could have treatments that would radically extend the life span by 2020 if we invested in a research program significantly cheaper than the NNI or the Bush administration's proposed mission to the Moon and Mars. Instead, under the Bush administration, U.S. research investments have shifted into areas with military and defense applications.

Support appropriate regulations on scientific research and
technological innovation

Democratic transhumanists should promote rigorous, independent safety testing of transhuman technologies, rejecting both free-market laissez-faire and Luddite bans. International agencies should be empowered to enforce global regulations on the safety of industrial and medical technologies. The U.S. Congress should reestablish the Office of Technology Assessment, and the size and mandate of the EPA and FDA should be expanded to rapidly vet the safety of new industrial materials, drugs and medical devices. Executive bioethics advisory bodies should be appointed that are representative of the bioethics community, and not stacked to produce preordained bioLuddite outcomes.

Protect genetic self-ownership and the genetic and intellectual commons from patent madness

Patents on existing genomes of plants, animals and humans should be declared void. Patents on novel gene sequences should be protected, unless they end up as part of the body of a self-aware citizen, in which case that person should become co-owner of that genetic information. Individuals must have control over their own genome, extending to the privacy of their genetic information.

DEFEND AND EXTEND SOCIAL RIGHTS

Build and defend universal health systems with choices

All citizens should be guaranteed equitable access to a basic package of health care services, including enhancement technologies when fiscally possible. When safe enhancement technologies cannot be provided through the public health system for political or fiscal reasons, they should be available in the marketplace.

Establish a guaranteed basic income and expand the social wage

All citizens should be guaranteed a basic income. Public financing of higher education should be expanded.

CREATE GLOBAL SOLUTIONS

There is no salvation for civilization, or even the human race, other than the creation of a world government. With all my heart I believe that the world's present system of sovereign nations can lead only to barbarism, war and in-

humanity, and that only world law can assure progress toward a civilized peaceful humanity.

—ALBERT EINSTEIN

It took man 250,000 years to transcend the hunting pack. It will not take him so long to transcend the nation.

—J.B.S. HALDANE,
"Daedalus, or Science and the Future"

Just as human enhancement is the next step for individual evolution, global governance is the next step for human social evolution. In fact, global governance may be a necessary condition for the successful building of transhuman society. As technology advances, so do the risks from new weapons and accidents of mass destruction. Technologies of human enhancement and artificial intelligence magnify those risks and benefits and make the case for strengthening transnational governance even more urgent than it was in the era of nuclear weapons and germ warfare.

Build democratic global governance

The growing integration of the European Union is proof that a global evolution to federalism may be slow and messy but inevitable. It is not a coincidence that the European Union is trying to promote the United Nations and multilateralism to restrain Pax Americana. We need global agreements not just to expand "free trade," but also to protect worker rights and set environmental and safety standards for agriculture, industry and medicine. The United Nations needs the authority to tax corporations and nations, and the power to collect those taxes. We need to add a second chamber to the United Nations that represents the world on a population basis, not just as nation-states. We need a permanent, standing international army with a clear mandate to enforce world law, starting with the Universal Declaration of Human Rights.

Ensure access to technology for the developing world

Agencies in the developed world should expand research into technologies appropriate to the needs of the developing world, and support programs of technology transfer through the World Health Organization, Food and Agriculture Organization, United Nations Conference on Trade and Development, United Nations Development Programme and UNESCO.

Reduce global risks to the future of civilization

In his essay "Some Limits to Global Ecophagy by Biovorous Nanoreplicators, with Public Policy Recommendations," Robert Freitas suggests that the best way to ensure that rogue nanotechnology is not harming the ecosystem is to set up the kind of global ecosystem monitoring that we need in any case to monitor climate change, toxics, soil erosion and deforestation. The Federation of American Scientists says the best way to prepare for bioterrorism is to set up a global public health information system that will automatically detect the emergence of new diseases, a system we also need to deal with the far greater threat from naturally occurring diseases like ebola, cholera, TB, SARS and AIDS. The world needs international bodies like the International Atomic Energy Agency to be expanded into a global infrastructure of technological and industrial regulation capable of controlling the health and environmental risks from new technologies. We need to expand programs like the Intergovernmental Panel on Climate Change and the British and American programs monitoring near-Earth objects into global programs to monitor the health of ecosystems and the threat from asteroids.

In *Our Final Hour,* astrophysicist Martin Rees enumerates the various threats to the existence of the human race from natural phenomena like gamma-ray bursts and asteroids to technological disasters like runaway nanotech and ecological destruction. Rees gives humanity 50–50 odds of surviving the century and says the chance of extinction is all the more poignant since "the post-human potential is so immense that not even the most misanthropic amongst us would countenance its being foreclosed by human actions." It is precisely because we want to see a posthuman future that we have to begin to build institutions of global governance that ensure new technologies enhance and protect humanity, and do not threaten its future.

Further Resources

To make this book a little more affordable I have put the citations and a more complete appendix of further resources online at the Cyborg Democracy Web site:

http://cyborgdemocracy.net/citizencyborg.htm

In any case, nowadays you can find just about anything I mention in the text by "Googling" it. But in this section I do briefly want to suggest some additional readings that further illuminate the topics covered in each chapter.

CHAPTER 1: BETTER LIVING THROUGH SCIENCE AND DEMOCRACY

Acceleration and Convergence

The most important recent statement on the accelerating and converging technological trends and their impacts on radical human enhancement is the National Science Foundation's NBIC collection:

Roco, Mihail C., and William Sims Bainbridge, eds. 2002. *Converging Technologies for Improving Human Performance*. Washington, D.C.: National Science Foundation. http://wtec.org/ConvergingTechnologies

Ray Kurzweil is the leading proponent of the idea that accelerating technologies will lead to discontinuous social change and intimate nano-neural communication with computers within decades.

Kurzweil, Ray. 2000. "The Law of Accelerating Returns." http://www.kurzweilai.net/articles/art0134.html

Nanotechnology

Drexler, Eric. 1986. *The Engines of Creation: The Coming Era of Nanotechnology*. New York: Anchor Books. The ur-text of nanotech. http://foresight.org/EOC/index.html

The two leading think tanks addressing the transhuman implications of nanotechnology and nanomedicine are:

Foresight Institute (Palo Alto, California, USA) The base for Eric Drexler and a network of people working to bring about Drexler's vision of molecular manufacturing and nanobots. http://www.foresight.org

Center for Responsible Nanotechnology (New York, New York, USA) A group working to promote anticipatory public policies to ensure that molecular manufacturing is safe and widely available. http://www.crnano.org

CHAPTER 2: CONTROLLING THE BODY

Robert Freitas's series of *Nanomedicine* tomes are the place to start for would-be nanoengineers, but he has a full bibliography with links to shorter pieces that are more accessible introductions to some of the possible applications like robotic white and red blood cells. http://www.rfreitas.com/NanoPubls.htm

Coalition for the Advancement of Medical Research (CAMR) (Washington, D.C.) A lobby organized to defend stem cell research and regenerative medicine, composed of patient organizations, universities, scientific societies, foundations and individuals with life-threatening illnesses and disorders. http://www.stemcellfunding.org

Futurist and artist Natasha Vita-More has also pulled together a number of ideas about the posthuman body into an engaging Web presentation. http://www.natasha.cc/primointro.htm

An up-to-date review of the effect of inequality on longevity can be found in:

Marmot, Michael. 2004. *The Status Syndrome: How Social Standing Affects Our Health and Longevity.* New York: Times Books.

An important place to start for criticisms of germinal choice would be:

President's Council on Bioethics. 2003. "Chapter Two: Better Children," in *Beyond Therapy: Biotechnology and the Pursuit of Perfection.* http://www.bioethics.gov/reports/beyondtherapy/chapter2.html

CHAPTER 3: LIVING LONGER

The best way to stay in touch with radical life extension news is to subscribe to the daily or weekly e-mail bulletins from Betterhumans.com and peruse the work collected at these longevity-oriented sites:

Betterhumans.com A webzine devoted to human enhancement and longevity technology. http://betterhumans.com

Longevity Meme An activist organization working to promote radical life extension. http://longevitymeme.org

Immortality Institute An international network of people working on radical life extension. http://imminst.org

Methuselah Mouse Project An annual prize for the team that has made a mouse live the longest. http://methuselahmouse.org/

Some summaries of the lines of research that will lead to radical life extension, and a proposal for how to bring them all together, can be found in Aubrey de Grey's Web site for "SENS (Strategies for Engineered Negligible Senescence)." http://www.gen.cam.ac.uk/sens

There has also been a backlash from aging researchers against life extension hype. They don't buy the idea that accelerating, or even linear, progress will make life expectancies of more than 100 possible. Dozens of researchers signed on to a pessimistic statement, "The Truth of Human Aging," published in *Scientific American* in 2002, and then Stephen Hall took a whack at the field in *Merchants of Immortality*.

Olshansky, S. Jay, Leonard Hayflick and Bruce A. Carnes. 2002. "The Truth About Human Aging," *Scientific American*. May 13. http://www.sciam.com

Hall, Stephen S. 2003. *Merchants of Immortality: Chasing the Dream of Human Life Extension.* Boston: Houghton Mifflin.

The American Academy of Anti-Aging Medicine has a Web site devoted to rebutting these critics of anti-aging medicine. http://a4minfo.net/virtualinterview

Another place to start for criticisms of life extension would be:

President's Council on Bioethics. 2003. "Chapter Four: Ageless Bodies," in *Beyond Therapy: Biotechnology and the Pursuit of Perfection.* http://www.bioethics.gov/reports/beyondtherapy/chapter4.html

Cryonics

Ettinger, Robert. 1964. *The Prospect of Immortality.* New York: Doubleday. The book that launched the cryonics movement. http://www.cryonics.org/book1.html

Merkle, Ralph. 1994. "The Molecular Repair of the Brain," *Cryonics*, 15(1–2). Probably the most important article in boosting the field of cryonics. http://www.merkle.com/cryo/techFeas.html

CHAPTER 4: GETTING SMARTER

The best recent review of the Flynn effect, the evidence for rising intelligence in the industrialized world, is:

Neisser, Ulric, ed. 1998. *The Rising Curve: Long-Term Gains in IQ and Related Measures.* Washington, D.C.: American Psychological Association.

British behavioral geneticist Robert Plomin is one of the leading researchers whose work is suggesting a limited set of genes that influence intelligence. A place to start would be his recent article:

Plomin, Robert, and Frank M. Spinath. 2004. "Intelligence: Genetics, Genes, and Genomics," *Journal of Personality and Social Psychology*, January, 86(1).

CHAPTER 5: BEING HAPPIER

Dutch sociologist Ruut Veenhoven's World Database on Happiness Web site is at
http://www.eur.nl/fsw/research/happiness/

and it includes an extensive bibliography on the social correlates of subjective well-being.

For more on the genetically determined happiness set-point:

Lykken, David. 1999. *Happiness: What Studies on Twins Show Us About Nature, Nurture, and the Happiness Set Point.* New York: Golden Books Adult Publishing.

The principal transhumanist thinkpiece on hedonic self-determination is:

Pearce, David. 1998. "The Hedonistic Imperative" http://www.hedweb.com/hedethic/

Although it focuses on drugs that remove unhappy memories rather than happy drugs per se, an important place to start for criticisms of hedonic engineering would be:

President's Council on Bioethics. 2003. "Chapter Five: Happy Souls," in *Beyond Therapy: Biotechnology and the Pursuit of Perfection.* http://www.bioethics.gov/reports/beyondtherapy/chapter5.html

CHAPTER 6: FROM FUTURE SHOCK TO BIOPOLITICS

This fourfold schema of political ideology is derived from the empirical work done by William Maddox and Stuart Lilie, reported in their 1984 *Beyond Liberal and Conservative: Reassessing the Political Spectrum* (Washington, D.C.: Cato Institute).

CHAPTER 7: CYBORG CITIZENSHIP

Liminality, Anomie and Future Shock

The section starts with my reflections on Toffler:

Toffler, Alvin. 1970. *Future Shock.* New York: Bantam Books.

My reflections on the role of liminality and future shock in shaping the bioLuddite response are influenced by two classic anthropology texts:

Douglas, Mary. 1966. *Purity and Danger: An Analysis of the Concepts of Pollution and Taboo.* London and Henley: Routledge and Kegan Paul.

Turner, Victor. 1969. *The Ritual Process: Structure and Anti-Structure.* Chicago: Aldine.

Personhood Theory and Citizenship

A recent transhumanist argument for an expanded concept of legal personhood is found in Linda MacDonald Glenn's piece "Biotechnology at the Margins of Personhood: An Evolving Legal Paradigm." http://www.jetpress.org/volume13/glenn.html

Brain Death and Personhood

Gervais, Karen G. 1986. *Redefining Death*. New Haven: Yale University Press. The classic argument for a personhood-based definition of death.

Extrauterine Gestation and Fetal Personhood

Tooley, Michael. 1984. *Abortion and Infanticide*. Oxford: Oxford University Press. Classic argument for the non-personhood of the fetus and newborn.

Kuhse, Helga, and Peter Singer. 1985. *Should the Baby Live—The Problem of Handicapped Infants*. Oxford: Oxford University Press. Argues that the non-personhood of the newborn should warrant a greater flexibility for parents to make treatment decisions for handicapped newborns.

Bailey, Ron. 2003. "Babies in a Bottle," *Reason*, August 20. A recent review of technical progress toward extrauterine gestation, and its political repercussions. http://reason.com/rb/rb082003.shtml

Apes and Genetically Enhanced Animals

Cavalieri, Paolo, and Peter Singer, eds. 1993. *The Great Ape Project*. New York: St. Martins Griffin.

Wise, Steven M. 2002. *Drawing the Line: Science and the Case for Animal Rights*. New York: Perseus Books.

Keenan, Julian, Dean Falk and Gordon G. Gallup. 2003. *The Face in the Mirror: The Search for the Origins of Consciousness*. New York: Harper Collins. Summarizes research on the self-awareness of great apes, humans and other animals.

Posthumans and Machine Minds

McNally, Peter, and Sohail Inayatullah. 1988. "The Rights of Robots," *Whole Earth Review* (Summer): 2–10. Suggests that we will have to redefine our concept of rights to admit the possibility of robotic personhood. http://www.kurzweilai.net/articles/art0265.html

Solum, Lawrence B. 1992. "Legal Personhood for Artificial Intelligences," *North Carolina Law Review*, 70: 1231–1287. Considers whether made things are always property, and relates AI's potential personhood to bioethics. http://home.sandiego.edu/~lsolum/Westlaw/legalpersonhood.htm

CHAPTER 8: DEFENDERS OF NATURAL LAW

Religious Right BioLuddites

The Center for Bioethics and Culture and **The Council for Biotechnology Policy** (San Francisco, California, and St. Louis, Missouri, USA) led by Nigel Cameron. http://www.thecbc.org. http://www.biotechpolicy.com

The Center for Bioethics and Human Dignity (Chicago, Illinois, USA) led by conservative Christian bioethicist John Kilner, chair of ethics at Trinity International University. http://www.cbhd.org

The Biotechnology & American Democracy Program of the **Ethics and Public Policy Center** (Washington, D.C., USA) led by Eric Cohen, editor of *The New Atlantis*. http://www.eppc.org/programs/biotech

Some theological denunciations of the project of germinal choice and human enhancement include:

Lewis, C. S. 1943. *The Abolition of Man*. London: Macmillan. Includes a warning about genetic engineering. http://www.columbia.edu/cu/augustine/arch/lewis/abolition1.htm

Ramsey, Paul. 1975. *Fabricated Man: The Ethics of Genetic Control*. New Haven: Yale University Press. Methodist theologian Ramsey was one of the foundational writers in bioethics and a Christian opponent of germinal choice.

Congregation for the Doctrine of the Faith. 1987. "Respect for Life: Instruction on respect for human life in its origin and on the dignity of procreation" issued February 22, 1987. Vatican statement that condemns all genetic manipulation of the embryo since it likely involved invitro fertilization, doesn't respect the embryo and might lead to transgenic creatures. http://www.ewtn.com/library/curia/cdfhuman.htm

Council for Biotechnology Policy. 2002. "The Sanctity of Life in a Brave New World: A Manifesto on Biotechnology and Human Dignity." A manifesto, signed by many leaders of the U.S. Christian Right, stating that humans have dignity (souls) from conception to death, and that there should be an international treaty to forbid cloning and inheritable genetic modification to protect that dignity. http://www.thecbc.org/manifesto.html

Secular Conservative BioLuddites

President's Council on Bioethics (Washington, D.C., USA) chaired by Leon Kass; the many background documents and testimonies archived at the site are a comprehensive primer on the religious and secular opposition to human enhancement technology, which has been the PCB's principal focus under Kass. http://bioethics.gov

Fukuyama, Francis. 2002. *Our Posthuman Future: Consequences of the Biotechnology Revolution*. New York: Farrar, Straus, Giroux.

Kass, Leon. 1997. "The Wisdom of Repugnance," *The New Republic*, June 2. Kass's classic statement of his "yuck factor" philosophy. http://www.pbs.org/wgbh/pages/frontline/shows/fertility/readings/cloning.html

_____. 2002. *Life, Liberty and the Defense of Dignity: The Challenge for Bioethics*. San Francisco: Encounter Books. Kass's most recent statement of the need to stop life extension and human enhancement in order to protect human "dignity" from human liberty.

———. 2003. "Ageless Bodies, Happy Souls: Biotechnology and the Pursuit of Perfection," *The New Atlantis* (Spring). A critique of life extension. http://thenewatlantis.com/archive/1/kass.htm

Deep Ecological BioLuddites

Foundation on Economic Trends (Washington, D.C., USA) led by Jeremy Rifkin, and dedicated to the promotion of Jeremy Rifkin and his numerous projects to stop genetic research and technology. http://foet.org

International Center for Technology Assessment (Washington, D.C., USA) led by Andrew Kimbrell, former policy director for Jeremy Rifkin at FOET, and author of *The Human Body Shop*. Dedicated to a broad anti-technology agenda. http://www.icta.org/aboutus

Foundation for Deep Ecology (Sausalito, California, USA), founded and funded by millionaire turned eco-anarchist Douglas Tompkins. http://www.deepecology.org

Some of the key deep ecological bioLuddite texts are:

Devall, Bill, and George Sessions. 1984. *Deep Ecology: Living as If Nature Mattered*. Salt Lake City: Peregrine Smith Books. The classic statement of deep ecology philosophy.

Kaczynski, Ted. 1996. *The Unabomber Manifesto*. Kaczynski, known as the Unabomber when on his terrorist bombing spree, specifically addresses the alleged threat to freedom from human genetic engineering, and insists that the only way to stop it is to destroy industrial civilization including modern medicine. http://www.unabombertrial.com/manifesto/

Kimbrell, Andrew. 1998. *The Human Body Shop: The Cloning, Engineering, and Marketing of Life*. Washington, D.C.: Regnery Publishing. Kimbrell's most recent statement on the threats from biomedical capitalism.

Joy, Bill. 2000. "Why the Future Doesn't Need Us," *Wired*, April. Although he is not a deep ecologist, Joy's article is a classic argument for technology bans on the basis of potential apocalyptic risks. http://www.wired.com/wired/archive/8.04/joy.html

McKibben, Bill. 2003. *Enough: Staying Human in an Engineered Age*. New York: Times Books. His title ably sums up the argument of the book—he's satisfied and doesn't see any point in making things better with technology, especially if it means becoming posthuman.

CHAPTER 9: LEFT-WING BIOLUDDITES

Society for Genetics and Society (Oakland, California, USA) led by longtime leftist activist Marcy Darnovsky and the former assistant political director for the

Sierra Club Richard Hayes. Dedicated to stopping "techno-eugenics" and banning cloning and inheritable genetic modification. http://genetics-and-society.org

Council for Responsible Genetics (Boston, Massachusetts, USA), once focused on genetic patents and public safety issues, but now led by a Rifkinite board that has signed on to a broader anti-biomedical agenda including opposition to stem cells and therapeutic cloning. http://www.gene-watch.org

ETC Group, The Action Group on Erosion, Technology and Corporate Concentration (Winnipeg, Canada), formerly RAFI, is an international organization headquartered in Canada, working on development from an anti-corporate perspective. Recently they have focused attention on the threats posed by human genetic engineering and nanotechnology. http://www.etcgroup.org

Some key left-wing bioLuddite texts are:

Rifkin, Jeremy. 1983. *Algeny*. New York: Penguin. The book that defined Rifkin's bioLuddism.

Rothman, Barbara Katz. 1988. *Recreating Motherhood: Ideology and Technology in Patriarchal Society*. New York: Norton. Argues that reproductive technology combines the evils of capitalism, patriarchy and technology.

Council for Responsible Genetics. 1992 (2000). "Position Paper on Human Germline Manipulation." http://www.gene-watch.org/programs/cloning/germline-position.html

Hubbard, Ruth, and Elijah Wald. 1997. *Exploding the Gene Myth*. Boston: Beacon Press. Argues that genetics won't accomplish as much as people expect, and that the focus on genetic therapy distracts from social problems.

World Watch Magazine. Special Issue: Beyond Cloning. July 2002. A dozen essays attacking human enhancement from a Left point of view. http://www.worldwatch.org/pubs/mag/2002/154

Winner, Langdon. 2003. "Are Humans Obsolete?" An extended critique of the misanthropy and elitism of the libertarian transhumanists, and a call for a more left-wing technopolitics, open to technological progress but sensitive to social reform and inequality. http://www.rpi.edu/~winner/AreHumansObsolete.html

ETC Group. 2003. *The Strategy for Converging Technologies: The Little BANG Theory*. Winnipeg: ETC Group. Argues that innovation on human enhancement should be stopped until the world is made more equal. http://www.etcgroup.org/article.asp?newsid=378

Mehlman, Maxwell J. 2003. *Genetic Enhancement and the Future of Society*. Bloomington: Indiana University Press. Argues for a global ban on genetic enhancement to prevent a genetic caste system.

Elliot, Carl. 2003. *Better Than Well: American Medicine Meets the American Dream*. New York: Norton. Desires for enhancement are capitalist-induced mental illness.

CHAPTER 10: UPWINGERS, EXTROPIANS AND TRANSHUMANISTS

Renaissance and Enlightenment Transhumanism

Mirandola, Giovanni Pico della. 1486. "Oration on the Dignity of Man." http://www.santafe.edu/~shalizi/Mirandola

Bacon, Francis. 1620. *Novum Organum.* http://history.hanover.edu/texts/Bacon/novorg.html

Early Twentieth-Century Transhumanism

Wells, H. G. 1898. *The Time Machine.* A classic dystopian reflection on how class inequality might lead to subspeciation of the human race. http://www.bartleby.com/1000

Haldane, J. B. S. 1924. *Daedalus, or Science and the Future.* New York: E. P. Dutton & Co. The classic essay by one of the founders of modern genetics on the future use of extrauterine gestation and other technologies, which inspired Huxley's *Brave New World.* http://www.santafe.edu/~shalizi/Daedalus.html

Bernal, J. D. 1929. *The World, the Flesh & the Devil: An Enquiry into the Future of the Three Enemies of the Rational Soul.* Speculates about space colonization, bionic implants and mental improvements. http://www.santafe.edu/~shalizi/Bernal

Stapledon, Olaf. 1930. *Last and First Men.* London: Methuen. A boring book that inspired many with its visions of future posthuman possibilities.

Huxley, Aldous. 1932. *Brave New World.* Garden City, N.Y.: Doubleday, Duran & Co. The classic anti-transhumanist critique of the dehumanizing and authoritarian possibilities of transhuman technologies such as mood drugs and extrauterine gestation. http://somaweb.org

Post–World War Two Transhumanism

Ettinger, Robert. 1972. *Man into Superman.* New York: St. Martin's. Ettinger's vision of transhumanity. http://www.cryonics.org/book2.html

Sterling, Bruce. 1985. *Schismatrix.* New York: Arbor House. A novel that imagines transhumanism as a political philosophy uniting a diverse posthumanity.

FM–2030. 1989. *Are You a Transhuman?* New York: Warner Books.

Regis, Ed. 1990. *Great Mambo Chicken and the Transhuman Condition.* New York: Perseus Books. An upbeat gonzo journalistic account of some of the pre-extropian transhumanists.

Alexander, Brian. 2003. *Rapture: How Biotech Became the New Religion.* New York: Basic Books. Focuses on the parallel growth of the transhumanist subculture and the gradual realization of biotech pioneers that the biotech revolution might make radical life extension possible. Ends before the emergence of the WTA.

Human Evolution, Reproductive Technology and Germinal Choice

Fletcher, Joseph. 1974. *The Ethics of Genetic Control.* New York: Doubleday Anchor. Fletcher made some of the first arguments for genetic choice, although he included a very problematic side argument for the creation of subhuman servants.

Glover, Jonathan. 1984. *What Sort of People Should There Be?: Genetic Engineering, Brain Control and Their Impact on Our Future World.* New York: Penguin. A defense of human genetic engineering that makes the distinction between therapy and enhancement, but accepts that enhancement could be ethical.

Robertson, John. 1994. *Children of Choice: Freedom and the New Reproductive Technologies.* Princeton: Princeton University Press. One of the most important statements of the case for procreative liberty, stopping short of germinal choice.

Hughes, James J. 1996. "Embracing Change with All Four Arms: A Post-Humanist Defense of Genetic Engineering," *Eubios Journal of Asian and International Bioethics,* June, 6(4). http://www.changesurfer.com/Hlth/Genetech.html

Silver, Lee. 1998. *Remaking Eden.* New York: Avon. Transhumanist and upbeat, but endlessly cited for its prediction of the emergence of GenRich and GenPoor.

Pence, Greg. 1998. *Who's Afraid of Human Cloning?* Boulder: Rowman and Littlefield. A defense of reproductive cloning.

Savulescu, Julian. 2001. "Procreative Beneficence: Why We Should Select the Best Children," *Bioethics,* 15(5/6). Argues that parents have an affirmative obligation to select the child, among the possible children they could have, who will have the best life. Includes a defense of selecting nondisease genes. (Available online at a long URL—Google it)

Stock, Gregory. 2002. *Redesigning Humans: Choosing Our Genes, Changing Our Future.* New York: Mariner Books. The latest and most popular defense of germinal choice.

Baylis, Francoise, and Jason Scott Robert. 2004. "The Inevitability of Genetic Enhancement Technologies," *Bioethics* 18(1): 1–26.

Extropians and Libertarian Transhumanism

Extropy Institute (Austin, Texas, USA) The main network for libertarian transhumanists, although its leadership insists that it is not libertarian. http://www.extropy.org

Friedman, David. 1989. *The Machinery of Freedom: A Guide to Radical Capitalism.* La Salle, Ill.: Open Court. An argument for replacing governments with "privately produced law," recommended by the Extropy Institute.

More, Max. 1990. "Deep Anarchy: An Eliminativist View of the State," *Extropy,* 5 (Winter). http://www.mind-trek.com/articles/t21c.htm

_____. 2000. "The Extropian Principles, Version 3.0: A Transhumanist Declaration." http://www.extropy.org/principles.htm

Hanson, Robin. 1994. "What if Uploads Come First: The Crack of a Future Dawn," *Extropy*, 6(2). Marxian immiseration in The Matrix. http://hanson.gmu.edu/uploads.html

———. 1998. "Is a Singularity Just Around the Corner? What It Takes to Get Explosive Economic Growth," *Journal of Evolution and Technology*, 2. http://www.jetpress.org/volume2/singularity.htm

Burch, Greg. 2001. "Progress, Counter-Progress, and Counter-Counter-Progress," presented at Extro 5, June 16. http://gregburch.net/progress.html

Bailey, Ron. 2004. *Liberation Biology: An Ethical and Scientific Defense of the Biotech Revolution*. Amherst, N.Y.: Prometheus Books.

The Grey Future: Robotics, AI, Singularity and Uploading

Clynes, Manfred, and Nathan S. Kline. 1960. "Cyborgs and Space," *Astronautics*, September, pp. 26–27, 74–75. The seminal essay for NASA on building a human being integrated with technology that regulates their body and emotions. http://search.nytimes.com/library/cyber/surf/022697surf-cyborg.html

Bear, Greg. 1983. *Blood Music*. New York: Arbor House. The book that anticipated the idea of "gray goo" and nano–AI transcencion years before Drexler published *The Engines of Creation*.

Vinge, Vernor. 1993. "Technological Singularity," *Whole Earth Review*, 81 (Winter): 89–95. The essay that launched the term and concept of the "Singularity." http://www.ugcs.caltech.edu/~phoenix/vinge/vinge-sing.html

Bostrom, N. 1998. "How Long Before Superintelligence?" *International Journal of Futures Studies*, 2.

Kurzweil, Ray. 1999. *The Age of Spiritual Machines: When Computers Exceed Human Intelligence*. New York: Viking.

Moravec, Hans. 2000. *Robot: Mere Machine to Transcendent Mind*. Oxford: Oxford University Press.

Brooks, Rodney. 2002. *Flesh and Machines: How Robots Will Change Us*. New York: Pantheon.

Global Brain

Global Brain FAQ. A very useful summary of global brain theories. http://pespmc1.vub.ac.be/GBRAIFAQ.html

Stock, Gregory. 1993. *Metaman: The Merging of Humans and Machines into a Global Superorganism*. New York: Simon and Schuster.

Russell, Peter. 1995. *The Global Brain Awakens: Our Next Evolutionary Leap*. Palo Alto, Calif.: Global Brain, Inc.

Bloom, Howard. 2001. *Global Brain: The Evolution of Mass Mind from the Big Bang to the 21st Century*. New York: John Wiley & Sons.

Contemporary Transhumanism

World Transhumanist Association The world's leading transhumanist organization. http://transhumanism.org

BetterHumans The leading portal for transhumanist news and views. http://betterhumans.com

Bostrom, Nick, et al. 1998–2002. The Transhumanist Declaration. http://www.transhumanism.com/declaration.htm

———. 1998–2003. The Transhumanist FAQ 1.5. http://transhumanism.org/resources/faq15.doc

———. 2003. The Transhumanist FAQ 2.1. http://www.transhumanism.org/resources/faq.html

Bostrom, Nick. 2001. "Transhumanist Values." http://www.nickbostrom.com/tra/values.html

Naam, Ramez. 2004. *More Than Human: How Biotechnology Is Transforming Us and Why We Should Embrace It*. New York: Random House. An accessible introduction to the transhumanist perspective for the layperson.

CHAPTER 11: DEMOCRATIC TRANSHUMANISM

Critiques of Libertarian Techno-Utopianism and Elitism

Dery, Mark. 1996. *Escape Velocity: Cyberculture at the End of the Century*. New York: Grove Press. Focused on "body-loathing" in the arts.

Borsook, Paulina. 2000. *CyberSelfish: A Critical Romp Through the Terribly Libertarian Culture of High-Tech*. New York: Public Affairs.

Goertzel, Ben. 2003. *The Path to Posthumanity*. Goertzel, a serious AI researcher who is deeply involved in the AI/Singularity subculture, takes a friendly swipe at extropian misanthropy in this excellent online book in Chapter 15, "Extropian Elitism and Humanist Posthumanism." http://www.agiri.org/path

Techno-Feminism and Harawayan Cyborgology

Firestone, Shulamith. 1970. *The Dialectic of Sex: The Case for Feminist Revolution*. New York: William Morrow. Argues that women will never be free of patriarchy until artificial wombs free them from having to carry babies.

Piercy, Marge. 1976. *Woman on the Edge of Time*. New York: Knopf. Depicts a utopian society that uses artificial wombs and a dystopian society in which gender differences have been genetically enhanced to the point that women are enslaved.

Haraway, Donna. 1984. "The Ironic Dream of a Common Language for Women in the Integrated Circuit: Science, Technology, and Socialist Feminism in the 1980s or A Socialist Feminist Manifesto for Cyborgs," *Socialist Review*, 80: 66–80. The somewhat opaque essay that launched the field of cyborgological culture studies. It argues that feminists should reject a dualistic view that associates men with culture and technology and women with nature. Instead women should embrace the cyborg as a liberating transgressive icon and integrate the power of technology into themselves. http://www.rochester.edu/College/FS/Publications/HarawayCyborg.html

Gray, Chris Hables. 1995. *The Cyborg Handbook*. London: Routledge. A collection of essays on cyborgs, from Kline and Clynes through the cyborgologist culture critics.

———. 2001. *Cyborg Citizen: Politics in the Posthuman Age*. London: Routledge. A reflection on the relationship between technology and politics in the coming transhuman age.

Progressive Defenses of Germinal Choice and Reproductive Technology

Farquhar, Dion. 1996. *The Other Machine: Discourse and Reproductive Technologies*. London: Routledge. A feminist defense of reproductive technologies.

Wicker, Randolph. 2000. "The Queer Politics of Human Cloning," *Gay Today*, March 20. Randy Wicker is a veteran gay rights activist and founder of the Clone Rights United Front. He argues that gay rights and the right to use reproductive and genetic technology are intrinsically linked. http://gaytoday.badpuppy.com/garchive/tech/032000te.htm

Advocates of Universal Access to Enhancement Technologies

Resnick, David. 1997. "Genetic Engineering and Social Justice: A Rawlsian Approach," *Social Theory and Practice*, 23/3 (Fall): 427–448. Argues that genetic engineering can and should be used to promote social equality.

Dworkin, Ronald. 2000. *Sovereign Virtue: The Theory and Practice of Equality*. Cambridge, Mass.: Harvard University Press. Argues that the best way to develop the fullest liberty of citizens, which is the obligation of society, is to provide general equality in the resources necessary for equal opportunity, including universal access to genetic enhancement.

Buchanan, Allen, Dan W. Brock, Norman Daniels and Daniel Wikler. 2000. *From Chance to Choice: Genetics and Justice*. Cambridge, UK: Cambridge University Press. Four leading bioethicists thoroughly critique bioLuddite arguments against genetic enhancement, and argue for universal access to genetic enhancements.

Singer, Peter. 2002. "Shopping at the Genetic Supermarket." Argues that "the state should be directly involved in promoting genetic enhancement." http://www.petersingerlinks.com/supermarket.htm

Transhumanist Disability Advocates

Christopher Reeve Paralysis Foundation A powerful lobby for research into stem cell therapies and regenerative medicine to cure paralysis. http://www.christopherreeve.org

Technogaianism and Existential Threats
Anderson, Walter Truett. 1987. *To Govern Evolution: Further Adventures of the Political Animal*. Boston: Harcourt Brace Jovanovich. Anderson argues that we have no choice but to learn to democratically govern evolution, both of the planet and of the human race.

Freitas, Robert. 2000. "Some Limits to Global Ecophagy by Biovorous Nanoreplicators, with Public Policy Recommendations." How to prevent "gray goo." http://www.foresight.org/NanoRev/Ecophagy.html

Bostrom, Nick. 2002. "Existential Risks: Analyzing Human Extinction Scenarios," *Journal of Evolution and Technology*, 9. http://jetpress.org/volume9/risks.html

Mulhall, Douglas. 2002. *Our Molecular Future: How Nanotechnology, Biotechnology, Robotics, Genetics and Artificial Intelligence Will Transform Our World*. Amherst, N.Y.: Prometheus Books. A thorough statement of "nanoecology" and look at the ways molecular technology will help us prepare for and prevent natural and human-made disasters.

Rees, Martin. 2003. *Our Final Hour: A Scientist's Warning: How Terror, Error, and Environmental Disaster Threaten Humankind's Future in This Century—On Earth and Beyond*. New York: Basic Books. Rees lays out the various threats to the future of humankind, including human technologies, ecological destruction and cosmological catastrophes.

Arnall, Alexander. 2003. *Future Technologies, Today's Choices: Nanotechnology, Artificial Intelligence and Robotics*. Greenpeace. A nice summary, with surprisingly moderate policy conclusions. http://www.greenpeace.org.uk/MultimediaFiles/Live/FullReport/5886.pdf

CHAPTER 12: TRANSHUMAN DEMOCRACY

Regulating Germinal Choices
Bostrom, Nick. 2002. "Human Genetic Enhancements: A Transhumanist Perspective." http://www.nickbostrom.com/ethics/genetic.pdf

Patents on Human Genes
The resources on this topic are enormous, but here are two places to start:

Fishman, Rachel. 1989. "Patenting Human Beings: Do Sub-Human Creatures Deserve Constitutional Protection?" *American Journal of Law and Medicine*, 15(4): 461–482.

Canadian Biotechnology Advisory Committee. 2002. *Patenting of Higher Life Forms and Related Issues*. Report to the Government of Canada. http://biotech.gc.ca

Universal Voucher-Based Health Insurance
Bioethicist Ezekiel Emanuel presents the idea of a universal voucher system, with competing, radically different health plans, in his 1992 *The Ends of Human Life: Medical Ethics in a Liberal Polity* (Harvard University Press).

CHAPTER 13: DEFENDING THE FUTURE

Regulating and Engineering for More Cooperation

Walker, Mark. 2003. "Genetic Virtue." An argument for a conscious social policy to genetically enhance the moral virtues. http://www.permanentend.org/gvp.htm

Group Selectionism and Altruistic Genes

For more on group selectionist explanations of altruism and other social phenomena, see Elliott Sober and David Wilson's 1999 *Unto Others: The Evolution and Psychology of Unselfish Behavior* (Harvard University Press) and Alexander Field's 2001 *Altruistically Inclined? The Behavioral Sciences, Evolutionary Theory, and the Origins of Reciprocity* (University of Michigan Press).

CHAPTER 14: A SEXY, HIGH-TECH VISION FOR A RADICALLY DEMOCRATIC FUTURE

Building the Transhumanist Movement

Join the **World Transhumanist Association** at http://transhumanism.org

Building the Next Left

Get involved with your local radical democratic movement or party:

Socialist International The existing social democratic parties. http://www.socialistinternational.org

Progressive Caucus The Left wing of the U.S. Democratic Party. http://bernie.house.gov/pc/

My favorite text imagining a global transhumanist Next Left:

Wagar, Warren. 1992. *A Short History of the Future.* Chicago: University of Chicago Press. Wagar's socialist world government makes human enhancement technology universally available.

Fighting Body-Based Oppressions and Discrimination

National Organization of Women The leading feminist organization in the United States. http://now.org/

International Gay and Lesbian Human Rights Commission The leading international lobby for transgender, as well as gay and lesbian, rights. http://www.iglhrc.org/

National Transgender Advocacy Coalition The leading U.S. lobby for transgender rights. http://www.ntac.org

Disabled Peoples International A radical, global disability rights organization. http://www.dpi.org/

Anti-Racist Action Network A global network of anti-fascist activists. http://www.antiracist.org

Extending Rights to Intelligent Non-Humans
Great Ape Project The global campaign to extend human-level legal protections to the great apes. http://www.greatapeproject.org/

Center for the Expansion of Fundamental Rights Steven Wise's foundation working to promote legal personhood for animals with selfhood. http://www.cefr.org/

Expanding Rights to Control Our Body
NARAL Pro-Choice America (formerly the National Abortion Rights Action League) A leading lobby for reproductive rights in the United States. http://naral.org

International Planned Parenthood Foundation A leading global NGO working for reproductive rights in the developing world. http://www.ippf.org/

American Infertility Association A lobby for consumers of infertility medicine. http://www.americaninfertility.org

Expanding Rights to Control Our Mind
Center for Cognitive Liberty & Ethics A nonprofit working for freedom of thought and our right to control our own brains. Their issues include the right to use drugs and other brain technologies, and the right to brain privacy from technologies such as "brain fingerprinting." www.cognitiveliberty.org/

Drug Policy Alliance The leading drug policy reform organization in the United States. http://www.drugpolicy.org/

Democratizing Technological Innovation
Alliance for Aging Research Leading coalition working to support therapeutic cloning and stem cell research in the United States. http://agingresearch.org/

Computer Professionals for Social Responsibility Progressive computer scientists. http://www.cpsr.org

Federation of American Scientists A liberal science lobby. http://www.fas.org/

Genetic Alliance An international coalition of individuals, professionals and genetic support organizations working on genetic-based diseases, research, insurance, Medicaid, genetic support groups and disability resources. Opposed to genetic discrimination. http://www.geneticalliance.org

Union of Concerned Scientists An ecological and left-wing science lobby. Produced the "Scientific Integrity in Policymaking" report critical of the Bush administration in 2004. http://ucsusa.org/

Extending Access to Health Care Technology

Two leaders of the movement for a universal health care system in the United States are:

Physicians for a National Health Program. http://pnhp.org

Universal Health Care Action Network. http://www.uhcan.org/

Also see:

Program for Appropriate Technology in Health Works with NGOs and governments in the developing world to make health care technology more accessible. http://path.org

Guaranteeing a Basic Income

Basic Income European Network The coalition of European groups working to institute basic income guarantees. http:// www.basicincome.org

U.S. Basic Income Guarantee The network building support for a BIG in the United States. http://www.widerquist.com/usbig/

Thousands of texts arguing for a BIG going all the way back to Tom Paine's *Agrarian Justice* are available here: http://www.widerquist.com/usbig/books.html but one of my favorites is:

Hughes, James J. 2004. "Embrace the End of Work," *Betterhumans*, February 20. http://betterhumans.com/Features/Columns/Change_Surfing/

Creating Global Solutions

Citizens for Global Solutions (formerly the Campaign for UN Reform and World Federalist Association) The leading organization working to create global federal institutions. http://globalsolutions.org

World Social Forum The leading coalition of grassroots groups working to build democratic alternatives to neoliberal globalization. http://www. forum socialmundial.org.br

Index